2ND EDITION

BUILDING WITH SECONDHAND STUFF

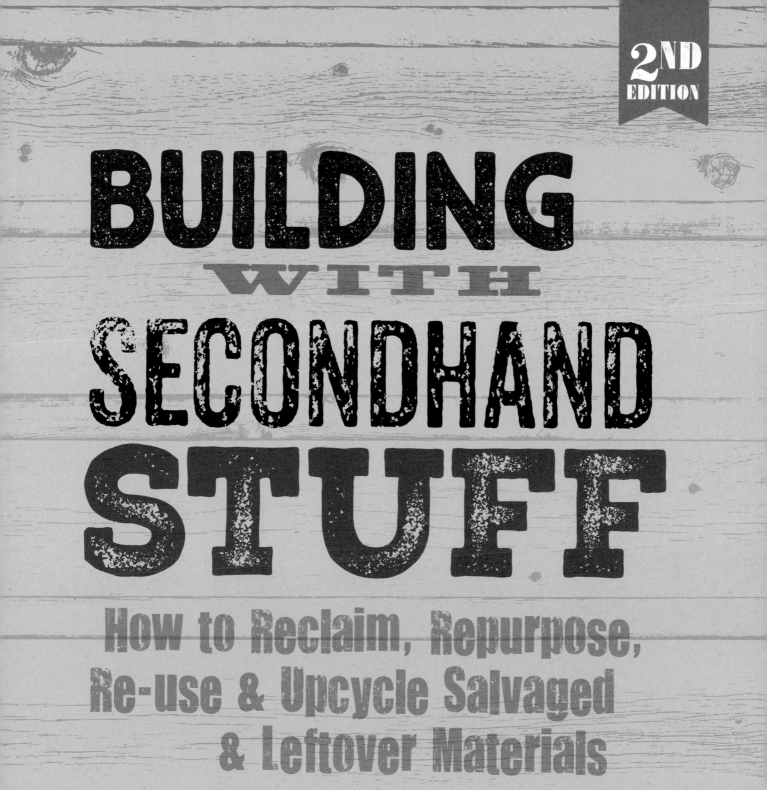

BUILDING
WITH
SECONDHAND
STUFF

How to Reclaim, Repurpose, Re-use & Upcycle Salvaged & Leftover Materials

CHRIS PETERSON

COOL
SPRINGS
PRESS
Home and Garden Experts™

MINNEAPOLIS, MINNESOTA

Quarto is the authority on a wide range of topics.
Quarto educates, entertains and enriches the lives of
our readers—enthusiasts and lovers of hands-on living.
www.quartoknows.com

First published in 2011 by Creative Publishing international, Inc., an imprint of Quarto Publishing Group USA Inc., 400 First Avenue North, Suite 400, Minneapolis, MN 55401 USA. This edition published 2017. Telephone: (612) 344-8100 Fax: (612) 344-8692

quartoknows.com
Visit our blogs at quartoknows.com

Cool Springs Press titles are also available at discounts in bulk quantity for industrial or sales-promotional use. For details contact the Special Sales Manager at Quarto Publishing Group USA Inc., 400 First Avenue North, Suite 400, Minneapolis, MN 55401 USA.

10 9 8 7 6 5 4 3 2 1

ISBN: 978-1-59186-681-7

Library of Congress Cataloging-in-Publication Data

Names: Peterson, Chris, 1961- author.
Title: Building with secondhand stuff : how to reclaim, repurpose, re-use &
   upcycle salvaged & leftover materials / Chris Peterson.
Description: 2nd edition ffl Minneapolis, MN, USA : Quarto Publishing Group USA
   Inc., 2017. ffl Includes index.
Identifiers: LCCN 2016046353 ffl ISBN 9781591866817 (paperback)
Subjects: LCSH: Waste products as building materials. ffl Construction and
   demolition debris--Recycling. ffl Building materials--Recycling. ffl Salvage
   (Waste, etc.)
Classification: LCC TA403.6 .P465 2017 ffl DDC 691--dc23
LC record available at https://lccn.loc.gov/2016046353

Acquiring Editor: Todd Berger
Project Manager: Alyssa Bluhm
Art Director: James Kegley
Layout: Kim Winscher

Printed in China

MIX
Paper from
responsible sources
FSC® C016973

CONTENTS

INTRODUCTION **7**

GALLERY OF SALVAGED BUILDING MATERIALS 11

CHAPTER 1:
WORKING WITH RECLAIMED MATERIALS 19

CHOOSING AMONG SALVAGED MATERIALS 20
SOURCING RECLAIMED MATERIALS 22
ASSESSING WHAT YOU FIND 24

CHAPTER 2:
RECLAIMING HERITAGE WOOD 27

RECLAIMED WOOD SPECIES 29
SALVAGING LUMBER 34
RECLAIMED WOOD TEXTURES 36
RECLAIMING WOOD 38
RECLAIMING WOOD TRIM & SIDING 39
RECLAIMED WOOD FINISHES 42
REMOVING SALVAGED WOOD FINISHES 43
STRIPPING PAINT 43
RECLAIMED WOOD REPAIRS 45
RECLAIMING WOOD FLOORING 47
RECLAIMING A VINTAGE WOOD FLOOR 47
RECLAIMING WOOD DOORS 48
RECLAIMED DOOR TYPES & STYLES 49
WORKING WITH RECLAIMED TIMBERS & BEAMS 53
REVIVING VINTAGE CABINETRY 56
RECLAIMING & REUSING PALLETS 57

CHAPTER 3:
SALVAGING OLD METAL 61

RECLAIMED TIN CEILINGS 64
RECLAIMING A TIN CEILING 65
CLEANING & PREPPING TIN PANELS 66
SALVAGED PLUMBING PIPE 67

CHAPTER 4:
RECLAIMING STONE MATERIALS 71

USES FOR RECLAIMED STONE & CERAMICS 72
WORKING WITH STONE TILES & SLABS 76
CLEANING STONE SURFACES 76
WORKING WITH LOOSE STONE 80
RECLAIMING KILN-FIRED MATERIALS 80
TYPES OF RECLAIMED BRICK 82
RECLAIMING & CLEANING BRICK 84

CHAPTER 5:
RECLAIMING GLASS **87**

REUSABLE OR NOT? 88
MIRROR IMAGE 89
RECLAIMED WINDOW TYPES 90
VINTAGE SPECIALTY GLASSES 95

CHAPTER 6:
SALVAGING FIXTURES &
ARCHITECTURAL ACCENTS **101**

TYPES OF SALVAGED FIXTURES & ACCENTS 102
RECONDITIONING RECLAIMED
    PLUMBING FIXTURES 106
CLEANING RECLAIMED HARDWARE 107
REMOVING PAINT 108
POLISHING & REFINISHING HARDWARE 108
SALVAGING VINTAGE LIGHTING FIXTURES 109

CHAPTER 7:
SECONDHAND STUFF PROJECTS **111**

HOW TO MILL TONGUES AND GROOVES INTO
    RECLAIMED PLANKS 112
HOW TO ROUTE TONGUE-AND-GROOVE FLOORING 114
HOW TO LAY TONGUE-AND-GROOVE RECLAIMED
    PLANK FLOORING 115
HOW TO SALVAGE PEGGED PLANK FLOORING 119
HOW TO LAY PEGGED PLANK FLOORING 119
END GRAIN FLOORING 120
HOW TO LAY RECLAIMED END GRAIN FLOORING 121
HOW TO HANG A RECLAIMED DOOR IN
    AN EXISTING JAMB 122
HOW TO BUILD A DOOR-BACKED ISLAND 128
PALLET PROJECTS 133
HOW TO MAKE A PALLET COFFEE TABLE 133
HOW TO CREATE A PALLET PLANTER 135
HOW TO INSTALL A TIN BACKSPLASH 136
HOW TO MAKE PIPE COAT HOOKS 139
HOW TO MAKE A PLUMBING PIPE TOWEL RACK 140
HOW TO BUILD A GRANITE-SLAB KITCHEN ISLAND 142
HOW TO BUILD AN OLD-WINDOW GREENHOUSE 146
HOW TO BUILD A RECLAIMED WINDOW CABINET 152
HOW TO BUILD AN ACCENT DOORKNOB COAT RACK 155

RESOURCES   **157**
INDEX   **158**

INTRODUCTION

America and Americans have long been defined by the ability to reinvent what came before and make something new out of the old. Everything we build is part of that process. Even the most sturdily constructed barns, homes, and commercial buildings come to the end of their lives and need to be taken apart so that something new can replace them. What's left is often a lot of building waste that fills already-strained landfills and dumps. But in the true American tradition, that waste is finding a new life, as thrifty and creative DIYers salvage, reclaim, clean up, and reuse the pieces of buildings no longer standing. This book is your guide to doing that—from identifying the best materials for your purposes, to cleaning and prepping them, all the way through integrating the materials into useful projects. This revised edition includes new projects ranging from the simple (a plumbing pipe towel rack) to the complex (an old-window greenhouse). To make it easier to get right to the projects, the projects are now organized in their own section. Select the one that speaks to you and start the process of turning trash into treasure.

Many—if not most—of the components that went into building structures that are ready to come down can be salvaged to find new life as reused home-design and building components. Using reclaimed building materials in new ways is part of a larger movement of sustainable building practices. In contrast to demolition, which sends tons of debris to landfills, this environmentally friendly process is called *deconstruction*. Deconstruction is the craft of taking apart the structure to preserve every piece that can be reused, repurposed, or recycled. It's one aspect of environmentally friendly living. But the rewards of reclaiming building materials and reusing them in your home extend far beyond helping the environment. That just happens to be arguably the most important benefit.

AN EASY GREEN SOLUTION

Landfills are filling up more quickly than we can find new places to put our garbage. We produce more waste per person than any other country on the planet. We're also consuming materials at an alarming rate. Old-growth forests have been overlogged for centuries, and now precious few are left. The US Environmental Protection Agency estimates that we throw away more than a billion board feet of lumber each year. If we reused all of that, we could save more than a million trees each year. There's simply no getting around that fact that much of the materials we've come to associate with quality in the home are either nonrenewable or very slow to renew. And those resources continue to be depleted.

Diminishing natural resources are not the only environmental repercussion of our ongoing hunt for

Building debris accounts for much of the waste clogging landfills—most of which could actually be put back into use.

Material	Relative Cost Savings	Reusables	Avoid	Safety Concerns
Wood	$$	Framing members Flooring/subflooring Siding/paneling Fencing	Pressure-treated Insect-damaged Water-damaged	Lead paint Cracked or compromised
Metal	$	Plumbing pipe (water) and joints Tin ceilings Architectural fittings	Lead pipe Rust	Lead Sharp edges
Glass	$$	Windows Period doorknobs Leaded lights	Broken panes	Non-tempered
Stone	$	Terracotta pots and tiles Pavers, curbstones Fieldstones Countertops (granite, soapstone, etc.)	Broken, sharp edges	Weight
Ceramic/porcelain	$	Tile Knobs, pulls Tubs and sinks	Broken, sharp edges	Lead paint

building materials. Finding, harvesting, and getting those materials to market means using vast quantities of fossil fuels and creating pollution. It's a cold, hard fact that the buildings we erect cost much more than we pay for them.

Fortunately, there are many solutions to these environmental issues. Increasingly, builders and contractors are using synthetic and recycled alternatives, or easily renewable resources such as bamboo. Home building and design are becoming more efficient, and other solutions will no doubt arise in the future. But a big answer to environmental impact and limited resources comes from the past. There is a treasure trove of usable materials in the buildings we tear down (or that fall down of their own accord). We just need to tap that resource.

LOWER COST, HIGHER QUALITY

Saving the environment could easily be a good enough justification to salvage and reuse building materials. But there are other great reasons as well—saving a little money is chief among them. The mini-industry that has sprung up to reclaim and resell used building materials competes with home centers and other retailers based largely on lower price and higher quality. Depending on whether you're shopping for heritage wood flooring or looking for a rescued slate slab for a countertop, you can expect to save from 10 to 50 percent off the price of comparable material sold brand new. In some instances, the owner of a building that must come down will let you take whatever you can safely remove, so the materials you salvage will cost nothing more than your labor.

Regardless of the source, you can expect to walk away with material that is as good or better quality than what you would pick up in the aisles of your local home center. Just about the only downside is that it may entail more research and work on your part, and much of what you find will not be in the standardized measurements common to building materials today.

Still, your wallet will thank you. And the money you save will be just part of your reward. Certain qualities and types of materials—especially woods—are no longer available. Rare types of quarry stone,

These Douglas fir timbers were rescued from a building about to be demolished. Now they'll be repurposed as mantels, counters, or even support beams, providing unique beauty and a tremendous bargain for lucky homeowners.

finely detailed architectural accents, and handcrafted leaded windows are just a sampling of the treasures that can't be matched in today's marketplace. In fact, much of the increasing popularity of reclaimed building materials is due to their unique nature and singular beauty. If you're in search of a one-of-a-kind look, reclaimed materials are a great place to start.

You may find rare species of woods not available, unusual stone surfaces with patterns and colors you're not likely to come across anywhere else, and glass that bears elegant imperfections from a less-standardized manufacturing process. Or maybe you're looking for something more modest, such as faucet or door handles that capture the charm of an earlier time, or wrought-iron grillwork that can be repurposed as an eye-catching fence or bench.

Whatever the reclaimed material, small design projects are a great place to start reusing them. Although building an entire home or large structure from reclaimed materials takes a great deal of dedication and planning, and requires specialized expertise, the projects we've collected for this book can be completed by any homeowner with modest DIY skills and a basic set of tools. We've purposely selected projects that are versatile and adaptable to any home. The techniques described here will provide a good base of skills and knowledge for working with any reclaimed materials, no matter what state they're in or what you want to do with them. Start with these projects and eventually you'll find even more ways to salvage and save.

GALLERY OF SALVAGED BUILDING MATERIALS

A mantelpiece made of reclaimed hand-hewn oak timbers is the perfect complement to a fireplace surround of reclaimed stacked fieldstone and a hearth of salvaged slate. The combination of elements is visually impressive and likely to last the life of the home and beyond.

Clad a kitchen oven column in reclaimed stone for a stunning design accent (and one that can easily withstand the heat extremes). The gray ledgestone shown here is ideal for this type of stacked application.

Reclaimed materials don't necessarily have to be reclaimed from defunct buildings. This gorgeous redwood table was crafted from a slab cut from a huge fallen tree. The tree was salvaged, having fallen naturally.

Above: *Pick a stunning window replacement at your local salvage shop. This stained-glass window features jaw-dropping painted accents and a design that would work in many different interior styles. It could transform a room, and probably for less than a new double insulated would run.*

Right: *Antique lighting fixtures are often unlike anything on the marketplace. Inside or out, they make for stunning—and functional—focal points.*

Keep an eye out for the unusual. One of the benefits of using reclaimed materials is uniqueness. This sculptural birdbath fountain needs a modest cleaning and it would be ready for reuse in a garden, where it would provide an entirely original focal-point feature.

Exploit the decorative potential of detailed historical accents. The handles and faceplates on display here are an example of the exquisite craftsmanship the building trades once produced. Introducing these handles or hardware into a contemporary home is a way to make a stunning visual statement with modest expense and little effort.

Above: Even hardware can be reclaimed for distinctive accents on kitchen cabinetry, doors, and windows.

Left: Look for small details that will supply big bang. Antique brass mortise latch sets like these can be used as accents in many different areas of the home and can even be displayed on their own.

Reclaim a cut slab for a truly impressive conversation piece in any room. Whether you reuse them as crude tables, rustic countertops, or a fireplace hearth like the one shown here, slabs bring a lot of visual power and a warm feeling to any interior.

Left: An antique doorknob is the only hardware that really looks right on an antique door. In most cases they can be used with modern door-latch mechanisms.

Opposite: Make a homey country kitchen even more inviting with the help of a wormy chestnut floor riddled with beautiful imperfections. No new floor can replicate the unique appeal and warmth of a plank surface like this.

Above: The saw patterns on this flooring, reclaimed from a granary, make it visually interesting and different from any modern version of oak planks. These planks would work equally well reused as flooring, wall paneling, or furniture.

Right: A mix of reclaimed teak and barnwood was used to build this rustic cabinet. The barnwood was left unfinished and the surface is a mix of textures and colors that invites the hands as well as the eyes.

CHAPTER 1

WORKING WITH RECLAIMED MATERIALS

Salvaging any usable building material—from barn wood to stained-glass windows to old pavers—requires some degree of cleanup and preparation for reuse. It might be as simple as washing decades of dirt off antique street pavers, or as complex as ripping down old tobacco barn wood for use as living room flooring. You can certainly leave that sort of work to the salvage company that rescues the material, if you don't mind paying more for the privilege. However, if you're handy with some basic tools and don't mind rolling up your sleeves and getting a little dirty, you can reclaim and revitalize secondhand materials as part of the process of reusing them. Not only will you save money, but you'll also have a chance to prep the material to your precise requirements. Common salvage techniques and tools are described in the pages that follow, along with the wealth of salvaged material options available to you.

Incredible diversity—not only among available materials, but between similar pieces as well—is at the heart of what makes salvaged materials so attractive. Time leaves its mark on all of us, but no more so than on the stone, wood, metal, and glass we build with. These materials age gracefully, featuring patinas of intriguing colors and patterns unlike anything you would find on newer products. Older building materials are often irregular, with imperfections that add to their allure. You'll be amazed at the different looks even among the same types of wood, or two similar bricks or stones.

Your search for the ideal building material that will bring your design project to life begins with the actual material you're considering. Whether it's a stunning antique door to replace a bland entryway, a new countertop, or a bedroom floor, you'll need to narrow down the myriad options. There are actually three different types of goods commonly salvaged from buildings: small architectural accents such as hooks, corbels, doorknobs, and hinges; single-piece fixtures including cabinet units, doors, and windows; and raw materials such as timbers, flooring, and stones.

CHOOSING AMONG SALVAGED MATERIALS

The reclaimed material you'll be looking for will of course be determined by your project. Some, such as a new fireplace mantel, can be crafted of different materials, from barn timbers to granite curbstones. Others, such as a living room floor, lend themselves to only one type of material.

And you shouldn't discount newer salvaged pieces. Some reclaimed building materials come from more recent buildings that are being extensively renovated and can present opportunities for intact building elements that are cheaper and just as good as new.

DECONSTRUCTION TOOLS

Square blade shovel (A); sledge hammer (B); hand maul (C); pry bars (D); claw hammer (E); "cat's paw" nail puller (F); utility knife (G); end-cut pliers (H); extractor pliers (I); locking-jaw pliers (J); metal detector (K); circuit tester (L); combination stud sensor/laser level (M); voltmeter (N); cordless drill bit assortments (O); pick axe (P); circular saw (Q); reciprocating saw (R)

WOOD

This is the most commonly salvaged material because it is so ubiquitous in both old and new construction. You can reuse wood from all the different areas of a building, including siding, structural timbers, existing wood flooring, and interior paneling. The wood can be used as-is, with the scars and marks of use in full display, or it can be resized and refinished to look brand new. Wood pieces can even be milled to serve new purposes, such as reconstituting solid-wood paneling boards into tongue-and-groove flooring. Old wood can find new life as tables, shelves, countertops, mantels, flooring, siding, and more. Entire wood structures, such as doors and shutters, are commonly resized and fitted for service in existing openings.

MASONRY

Many different types of stone and bricks are used inside and outside of buildings. Other types are used to pave roads and paths, and provide edging for sidewalks. All of these can be repurposed to stunning effect in a number of different areas of your home. Fireplace surrounds and mantels are extremely common uses for reclaimed stone. Depending on the architectural and decorative style of your house, cobblestones or certain pavers can make a lovely kitchen floor. Exclusive quarry stones such as granite, marble, and travertine can be used as stunning countertops or custom tabletops. Salvaged pavers and bricks are perhaps most at home outside, where they can be used to create a charming pathway through the garden or yard.

METAL

Metals such as iron and copper are regularly rescued from buildings being demolished. But because they are generally inexpensive and are so easily recycled, most salvaged metals find their way into the smelter. However, certain types of metal fixtures are well-suited to be repurposed as new home-design elements. Plumbing pipes and sheet forms, such as tin ceilings, are prime examples of metal pieces that can easily be adapted into a room design. You can cover an accent wall with hammered sheet metal or copper tiles for an arresting decorative feature. Use a heating vent grill to craft a coffee table, or make shelves or a pot hanger from copper pipes.

GLASS & CERAMICS

Most salvaged glass is reused for home design in the form of windows. The unusual shapes of some antique windows and the classic look of leaded-glass windows make those options particularly popular. But you don't have to limit yourself. Stained-glass designs are also regularly rescued, rehabilitated, and reinstalled in new locations.

SOURCING RECLAIMED MATERIALS

Once you've made the decision to use salvaged wood, stone, or other reclaimed building materials in a home-design project, the next step is to track down a good source. The bad news is that finding the perfect timbers for a trestle table, the right slab of marble to use for a new countertop, or a modestly sized leaded-glass window to put in that interior partition wall won't be quite as simple as heading out to the local home center. The good news is that you do have a lot of different resources to turn to in your search. And many of those represent the opportunity to negotiate an enviable price for your treasure.

Frankly, there are a lot of different sources for reclaimed building materials. Each has its benefits and drawbacks. The one you use will depend first and foremost on the type of material you're after. You'll also need to decide just how much work you're willing to do. Getting your hands on some materials can entail getting those hands a little dirty. You may be faced with the prospect of picking up the material if it's not local, or arranging for shipment. You may have to clean, revive, and prep the material before use. No matter what the case, always consider the full range of potential suppliers before making a decision.

RETAIL AND WHOLESALE SALVAGE COMPANIES

The past two decades have seen a steady growth in the number of companies reselling materials salvaged from demolished or renovated buildings. At the high-end are companies that deal primarily in architectural antiques, artifacts, and intact period structures such as decorative plaster corbels or hand-carved marble fireplace surrounds. The broader middle part of the

market is populated by general salvage firms dealing in raw building materials such as barn timbers and large curbstones. Some of these companies clean and prepare their inventory, while others simply reclaim and stock the materials, selling them as-is. Most usually specialize in a specific substance such as wood or stone, although many stock whatever materials come their way, and their inventories fluctuate accordingly. This is why looking for the material that's ideal for your project can be a little like a scavenger hunt. These companies should be your first stop. They're generally professional and easy to deal with. They can set up shipping if the need arises, and in most cases, you can rely on the integrity of the materials.

DEFUNCT STRUCTURES

If you have access to a dilapidated building—along with good DIY skills, knowledge of how buildings go together, and a solid selection of tools—you may choose to deconstruct and reclaim all or part of the building yourself. This can be a remarkably inexpensive way to recover building materials of all sorts—including hidden gems. You may, for example, find wood flooring in a beautiful antique species hidden under ratty old carpet. But local codes and regulations dictate safety practices (you should exercise appropriate safety measures in any case) and disposal methods that must be used in demolition or deconstruction. Some situations, such as the removal of asbestos, require extremely involved abatement procedures. Encountering toxic materials is one of the downsides to deconstruction, and abating material such as asbestos insulation can quickly escalate your costs.

CONTRACTORS AND DEMOLITION COMPANIES

Buildings don't necessarily have to be taken entirely apart for them to produce materials worthy of upcycling. Simple jobs such as a bathroom renovation can create piles of usable tiles, flooring, and pipes. A bigger home renovation can result in healthy quantities of antique wood flooring, cast-iron heating vent grills, or other desirables. Generally, however, it's not worth a contractor's time to carefully remove and resuscitate old tile or flooring. It's much easier to simply rip out the material and send it on a trip to the dump, never to be seen again. If you're willing to do a significant amount of work, you may be able to make a deal with a contractor: you provide the labor to demo the space, and in return, you keep the materials you remove. This is not an arrangement that suits everyone, but it can be a great way to lay your hands on unique building materials in exchange for sweat equity. In other cases, you may be able to work with large demolition companies to strip a building of valuable commodities before the company tears it down (or even as part of tearing it down). This type of arrangement works best when you need a large amount of wood, metal, or stone. It's not, however, for the uninitiated. Just as with deconstructing a building yourself, working in a job site space requires adhering to local regulations and codes and the same safety issues apply. You can make your life easier by limiting your agreement to combing through the roll-on dumpsters at the job site. The best time to get in on the process is when safety fences go up around a job site, or you see work permits appear in the windows of an older house.

ONLINE RESOURCES

The Internet has made selling small quantities of reclaimed building materials easy and simple. Even individuals can offer leftovers from renovations or building deconstructions, on websites such as Craigslist or eBay. Of course, buying from someone who isn't operating as a business and doesn't have a professional reputation to protect entails a certain amount of risk—especially if that source isn't local. You have to be cautious that you're getting the quality and specifications that the seller advertised. In-person inspection is always the best way to go. On the upside, materials sold online often offer incredible deals. Sometimes they are even offered free to anyone willing to haul them away.

GARAGE SALES, SWAP MEETS, AND FLEA MARKETS

This is the most hit-or-miss resource. But if you're willing to put in a fair amount of legwork, you can often stumble upon hidden treasures. Be prepared to haul away whatever you find, and come equipped with a measuring tape to determine if the dimensions of whatever it is will suit the project you have in mind. The informal nature of these types of venues lends itself to healthy negotiation and great deals. You can find notices of garage and yard sales in local newspapers— also a great place to look for individuals selling material through the classifieds.

ASSESSING WHAT YOU FIND

Once you've found a reliable source, the next step is to judge if the material itself is worth repurposing. Know the history of the donor building. This is especially important if you're looking at reclaimed wood or unsealed stone. It's just a basic health and safety issue. For instance, unfinished stone or wood used in an early twentieth-century factory may have been exposed to high levels of toxic chemicals. Much as you may love the look and price of the material, any stone or wood that was exposed to toxic or carcinogenic chemicals needs to be properly disposed of, not reused in a home.

Suffice it to say, it's well worth your while to ask questions and investigate as necessary to determine the history of whatever you're considering for reuse. Once you're certain the stone, wood, or metal isn't going to be a health problem, check the actual quality of the material. Judge the structural integrity, and look for obvious signs of damage, such as cracking, checking, or wear that would make reuse problematic. Check wood flooring for screw and nail holes too numerous to cover up with a little putty. Be diligent in looking for insect damage, even if it doesn't mar the appearance. For instance, wood timbers that look fine on the surface but have been structurally weakened throughout by termite damage are perhaps not a wise choice for a table or other load-bearing application. Unless you're willing to cut them down for reuse, cracked window glass and broken terracotta tiles should also be recycled, not repurposed.

In addition to basic structural integrity, consider the amount of preparation that will be required to make the material project-ready. For instance, old weathered redwood siding may look beaten up and too far gone to restore it to its original beauty, but a simple powerwashing can revive the appearance.

Aged bricks can be beautiful, colored with a timeworn patina that makes the surface of each one unique and fascinating. But if the bricks you've chosen are checkered with stubborn cement-based mortar, you may have considerable difficulty removing the mortar without destroying the brick.

SALVAGE WISDOM: RECLAIMING ARCHITECTURAL ACCENTS

No discussion of designing with reclaimed materials is complete without a mention of rescued architectural bits and pieces. These range from glass doorknobs that can add a lovely and elegant touch to doors throughout a home, to hinges and handles that can dress up plain wood cabinets, to much rarer, more special, and more expensive ornamentation such as plaster pediments and wooden finials. There are many companies that offer reclaimed architectural ornaments from both the US and Europe, although detailed fixtures such as decorative corbels can be quite pricey. However, in most cases, using small accents is one sure way to add big flair to your house without a big price tag.

CHAPTER 2

RECLAIMING HERITAGE WOOD

Wood, in all its many forms, is the most commonly salvaged building material. That should come as no surprise because it's one of the most attractive materials used in a home. It makes for an unrivaled surface underfoot, a warm and visually pleasant wall covering, a dramatic countertop choice, and so much more. Wood is also incredibly durable and forgiving, sturdy, and easy to work with. All of which is why most buildings feature wood framing. Some structures, such as barns and other utility buildings, are constructed solely of this compelling material. Start taking apart a building, or even a part of a building, and you'll be rewarded with a variety of wood pieces that you can reuse in a vast number of ways. Once you free it from its current location, you have only to decide which new use will best exploit the particular size, grain, texture, and color of the wood.

The grain pattern and color of most aged reclaimed wood is enthralling, because older lumber is wholly different from what you can find at a modern home center. Older reclaimed members are often different sizes than contemporary "nominal" lumber (an older 4 x 4 will actually be 4 inches x 4 inches) and may not be completely finished smooth on one or more sides. The surface may have been worked by hand with tools, or sawn in unusual ways, leaving unique surface textures. The aging process itself adds a unique look to many pieces of reclaimed wood. The differences can go even deeper, having to do with climate and logging.

Over the last few centuries, many old-growth forests have been logged out. The move toward conservation has led to managed forests full of new-growth trees. Instead of being cut from century-old behemoths, today's lumber is usually culled from relatively young growth. When a tree grows for hundreds of years, its basic structure becomes much more dense and stable. The wood of the tree has to hold it upright in the face of strong winds, inclement weather, and shifting terrain. Look at the end of an old-growth log and you'll see that the growth rings are much, much closer together than the rings of a newer tree. Because there was less carbon dioxide in the atmosphere, trees grew much more slowly hundreds of years ago. That slow growth created dense cellular structure. That density is why many old-growth barn timbers have held up for more than a hundred years.

This structure promoted the growth of *heartwood*. Heartwood is the hard, durable core of a large, old tree, the wood at its center that no longer carries sap or participates in bark growth. This part of the tree is the most desirable because it's the clearest wood with the fewest imperfections, and the strongest, hardest, and most durable. Once sawn into lumber, heartwood is also the most resistant to insects and water damage. It's usually the most uniquely colored and beautiful part of a tree, and often carries the scent of the tree. Because only trees of a certain age grow heartwood, you'll only find it in older buildings.

But that's not the only thing that makes reclaimed wood special. Buildings more than half a century old were often originally built of wood species that have now largely been logged out of existence or are rare enough not to be commonly used in construction. For instance, white oak and chestnut are stunningly beautiful woods that are no longer widely available. Find barn siding made of either of these, and you've lucked onto a true treasure.

RECLAIMED WOOD SPECIES

One of the principal attractions to reclaimed wood is the assortment of species available from old buildings and resellers alike. Some of these are rare enough that you may not have seen them before. But even more common species such as pine have different grain patterns and colorations than you'd find in modern lumber of the same species. Age has also left its mark on the wood, giving it a patina unlike any finish you'd see on samples of a more recent vintage. Chances are you'll be mesmerized by one or more of these captivating options. The species described on the following pages are the ones you're most likely to encounter in a hunt for woods to salvage. More exotic species such as mahogany, teak, and ipé haven't been regularly used in this country long enough to be reclaimed from older buildings.

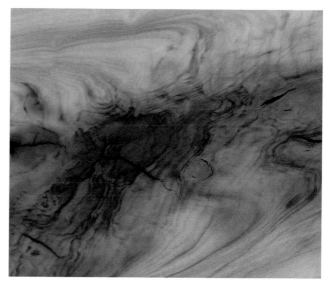

A close look at the surface of this reclaimed slab of bay laurel reveals the complex grain swirling, color variations, and imperfections that draw homeowners and others to antique woods.

RECLAIMED HARDWOODS

Wood	Profile	Appearance	Best Uses
American Chestnut	Often called "wormy" chestnut for the abundant wormholes. Trees nearly wiped out by blight in 1904. Wood weathers well.	Light brown to black. Imperfections are considered desirable and include checking, scarring, saw marks, and insect and nail holes.	Soft hardwood, making it easy to work with. Use in: flooring, paneling, and decorative accents. (Occasionally used in furniture.)
Antique Ash	Dense and strong enough to be the wood of choice for pro baseball bats.	Tight, long, oaklike grain. Creamy, almost-white sapwood most common. Rarer, darker brown heartwood is a better choice for floors.	Floors. Antique ash floors were often laid in varying widths for vibrant effect.

Wood	Profile	Appearance	Best Uses
Cherry	Favored by furniture makers, wood is quite beautiful and elegant, but supply is limited.	Rich red to reddish brown, deepening with age. Elegantly uniform, fine grain with small, charming imperfections.	Furniture, shelves, and plank flooring.
Antique Elm	US elm population has been devastated by Dutch Elm disease, so usually only found as salvaged or reclaimed wood.	Understated look, from tan to red-tinted brown, with simple flowing grain.	Traditionally used in furniture (and caskets). Best reused in furniture, paneling, and flooring.
Hickory	Fruit-bearing variety of pecan tree, one of the hardest and toughest woods, and can easily last the life of your home.	Pale, nearly white sapwood; dark, reddish-brown heartwood. Tight and straight grain, and exceptionally beautiful wood. Coarse textured.	Traditionally used in tool handles, exceptional when reused as incredibly durable flooring.
Maple	Long the wood of choice for high-end homes across the country, the wood comes from the same tree that yields maple syrup. Dense and strong wood.	Deep, creamy white color, accented with a fine, flowing grain. Some species feature curly grains, while bird's-eye maple offers intriguing circular grain pattern.	Has been used as shoe heels, cutting boards, and bowling alleys, but mostly used as furniture, shelving, and accents throughout time—and can be reused for these (as well as durable flooring).

Wood	Profile	Appearance	Best Uses
Antique Oak	Similar to the contemporary version, but with more visual variations in grain and color.	Pronounced grain, deep honey tones, from gold to rich milk-chocolate brown. Irregular swirls sometimes called Tiger Stripes. Older oak often has knots and other attractive imperfections.	Most often used as flooring in plank or strip form (reclaimed oak may need to be ripped down to avoid damaged sections), less often used as paneling.
Walnut	A traditional furniture wood, walnut is fairly rare but worth the hunt. Rare in older buildings, it's most often reclaimed from old barns built from local trees.	Heartwood is a sumptuous, deep almost-chocolate brown; the sapwood is lighter brown. Colors soften and meld with age; grain patterns range from straight to wavy.	Most sought-after walnut is black walnut, but all walnut makes incredible furniture, flooring, paneling, cabinet cases, and doors.
White Oak	A handsome wood, white oak is not considered as valuable as walnut or cherry, but does offer a nice look.	Light beige-brown to tan, with tight graining. The look is simple and elegant if not showy.	Historically used in casks and barrels, white oak is best reused in cabinets, paneling, and flooring.

Wood	Profile	Appearance	Best Uses
Cedar	A useful moisture- and insect-resistant wood that can be salvaged from many homes and structures.	Attractive fine grain and fairly uniform coloring ranging from dark straw to deep yellow-tan.	Decking (porch or deck), ceiling, and wall paneling.
Cypress	An unusual wood with the tough character of a hardwood thanks to abundant natural oils.	Dark appearance from mid-to deep brown; exceptional flowing grain with intermittent dark marks adding interest.	Originally used indoors and out, the wood serves well in moist environments, as flooring, cabinets, shelves, and even support beams.
Douglas Fir	A species of pine that is a hard softwood, sturdy and strong. Often used in structural members.	Rustic appeal with a loose, irregular grain that gets tighter with age. Ranges from light to dark tan, aging to amber.	Good for ceiling and wall paneling, molding, flooring, and shelves. Easy to work with and takes stain and other finishes well.

Wood	Profile	Appearance	Best Uses
Hemlock	Similar in look and durability to Douglas fir, hemlock is a hard softwood that was historically used in barns.	Uniform light yellow-tan that ages well, especially when sealed. Vivid grain patterns.	Flooring.
Pine	Reclaimed older pine is usually desirable heartwood, and the wood is so historically popular that it's some of the most widely available.	Wide, flowing grain patterns; attractive yellow to golden color that ages to amber.	Flooring, shelving, paneling, and wainscoting.
Redwood	Classic wood for decks, rustic interiors, and outdoor furnishings.	The tight, even grain and alluring red-brown color account for redwood's timeless appeal. Unsealed, it weathers to an elegant gray.	Light wood that is easy to work with, redwood is reused as wood paneling, railing, furniture, and countertops.

SALVAGING LUMBER

The process of salvaging wood is not difficult; you'll find potential treasure troves of usable material from buildings under deconstruction to construction-site dumpsters. The wood you're looking for will usually be determined by what it is you want to build. But keep your eyes open and be a little flexible in your search. Oversized lumber can easily be milled to dimensions that work for you.

For instance, planks from siding are often milled with tongues and grooves so that they can be used as flooring. Use the same process to rip plank flooring into strips. Looking for 4×4s for a chunky, eye-catching trestle table? If you can't find just the right pieces, laminate reclaimed 2×4s together instead. Keep in mind that it's hard to mix reclaimed and new lumber in the same project, because older members may be actual, rather than nominal, sizes.

RECLAIMED LUMBER TYPES

Wood	Profile	Appearance	Best Uses
Timbers & Beams	The most impressive reclaimed wood lumber; large structural timbers rescued from old barns, warehouses, and similar structures.	Visually fascinating, scarred with the marks of instruments used to square them off. Timbers can run as large as 2 feet square and up to 20 feet long. Headers 4 to 6 inches thick are common.	Can be used in their original form for a rough, rustic look. Or can be resurfaced and refinished as sleek, exposed structural elements. Smaller beams are used for mantels, shelves, and furnishings. Framing members in distinctive woods can be fabricated and finished to create remarkable countertops.
Planks	Reclaimed from floors of large, old structures such as warehouses, and the sides of old homes, barns, and rural outbuildings.	Wildly varied. If used as floors, may have an existing finish that can be rescued. Siding may bear colorations from exposure to the elements that can be stripped or left partially intact under a protective finish.	Planks are most often squared off or milled shiplap. But plain planks can be milled with tongues and grooves for use as flooring or paneling. They can also be ripped to produce strip flooring.
Flooring	Much of the usable wood from old buildings is flooring. Most is plank flooring, the more common style prior to mid-twentieth century.	May be tongue-and-groove, shiplapped, or butt-joined (shiplap is more common, the older the building). Older wide planks are usually longer than contemporary strips, 8 feet or longer.	Squared planks, such as siding, are usually milled with tongue and groove before being used as flooring. Leave as planks if possible because the wider surface allows more grain pattern and color to show.

Wood	Profile	Appearance	Best Uses
Doors	Reclaimed doors add wonderful flair to any opening in the house, from kitchen to the front entryway. Salvaged units are usually solid wood.	Come in the same fascinating grain patterns and colorations available in planks, timbers, and flooring. Door styles are nearly limitless, with arched single or double doors a common find. Doors to fit standard openings are available in every panel configuration imaginable.	Doors are usually repurposed as doors, but simple, solid types are sometimes upcycled into desks or bed headboards or footboards. Shutters of all shapes and sizes are an intriguing offshoot that can be adapted for use as cabinet doors or decoration. Always choose a door close to the final measurements of the opening.

SALVAGE WISDOM: TOBACCO BARN WOOD

Wood used in antique tobacco-curing barns is some of the most unique you'll ever find. Traditional curing barns were built from the eighteenth through mid-twentieth centuries. Modern structures replaced them starting in the 1970s, making the barns obsolete. The barn wood was usually of mixed species, a result of being built from whatever local trees were available. Any given barn may include antique pine, hemlock, cypress, or other woods. The curing process involved maintaining charcoal fires at temperatures above 100°F. Over time, the hot smoke impregnated the wood, often saturating it to as much as 1/2" below the surface. This created a range of intense, captivating colors in the wood, colorations that were different board to board and barn to barn.

Resellers deconstruct the barns and mill the boards just enough to create smooth and square faces and edges. They sometimes add tongues and grooves to create flooring and paneling. Because the structures are exceedingly rare, resellers are your best source for tobacco barn wood.

RECLAIMED WOOD TEXTURES

Surface texture is another of the many unique qualities that separate reclaimed wood from new lumber. Some older lumber will have a perfectly smooth surface. But you'll also find many other textures on the wood you salvage or find at a reseller. Older machinery and wood-processing techniques often left the wood's surface marked in unusual ways. The time, effort, and expense necessary to completely finish the surface were usually prohibitive or considered unnecessary. These days, those manufacturing marks are valued as evidence of history and character.

ROUGH SAWN

Rough sawing is the first stage of several in milling a piece of lumber. Wood mills have historically left certain support timbers and other structural members rough sawn to save time and money. This was especially true in the lumber used to construct barns, warehouses, and other purely functional buildings, where precise measurements and polished appearances didn't matter. Both hardwoods and softwoods were left rough sawn. Today, the coarse look and rough texture of a rough-sawn piece of lumber can lend a novel decorative element to a home, as long as it's used thoughtfully. Rough-sawn planks make wonderful rustic paneling in a den, and rough-sawn timbers provide a remarkable focal point when incorporated as exposed support beams, posts, or headers. In any case, the texture is best used sparingly in home design and it should be left unfinished and unpainted.

CIRCLE SAWN

Circle saws were used in mills before the advent of modern mill machinery. The process leaves regularly spaced, clearly apparent arcs across the face of cut boards. Circle saw marks are considered handsome proof of age and make for an extraordinary look in paneling, tables, and flooring. Circle-sawn lumber can be lightly sanded for refinishing without diminishing the impact of the saw marks.

SKIP SAWN

A predecessor to circle sawing, skip sawing creates much the same marks as you would find on a circle-sawn board, except that the marks are more random. It is a more informal and rustic look, but used in the right situation, it can be a very attractive surface texture for floors or wall paneling.

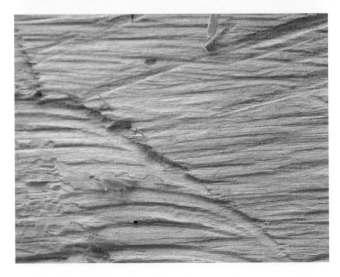

HAND HEWN

This term refers to timbers cut by hand with an adze or axe, and worked with other tools to create support beams and timbers. Found only in very old structures, hand-hewn timbers are usually hardwoods or softwoods from trees felled on the site, and are often rare species. The tool marks and imperfect surfaces of these timbers are considered charming accents to the wood's grain and color, and the timbers are usually used as exposed beams or mantels after a light resurfacing.

SMOOTH PLANED

In some cases, you may want to focus entirely on the natural grain and color of the wood, and not the distress marks of age. Smooth-planed boards and timbers are, as the name indicates, surfaced smooth to the touch, removing all apparent imperfections and completely revealing the natural character of the wood. But even in this case, the finish may have mellowed with time, offering captivating rich color and tone. That's why the finish is often preserved on smooth-planed members.

SALVAGE WISDOM: THE QUARTER-SAWN ADVANTAGE

Not to be confused with sawing techniques that texture the face of the wood, quarter sawing is a method of cutting logs into boards. Where the more common plain sawing cuts boards out of the log in parallel layers, quarter sawing cuts boards at 90° angles to the growth rings, essentially in a pattern radiating out from the center. This yields fewer boards and greater waste than plain sawing does. But the grain of quarter-sawn boards is much more captivating. Straight and tight grain lines usually run the length of the board, with small rays and fleck, imperfections scattered throughout the surface pattern that make for a very attractive surface look. Quarter-sawn boards are also considered better lumber, because the structure resists warping, twisting, and other distortions. The grain produced is also more water resistant, although quarter-sawn boards still take stain well.

RECLAIMING WOOD

Before you can use reclaimed wood in a home-design project, you need to actually reclaim the wood. You may find exactly what you need—and as much as you need—at a salvage company. But the reality is that inventories change with surprising frequency. This varying supply may mean that you have to look elsewhere for the wood you want. Or, it may just be that you want the hands-on experience of salvaging your own wood. Either way, it pays to know how to get the wood you need and prep it for the project you have in mind.

When you salvage wood from an existing building, the goal is to remove the lumber you want with minimum damage to the wood, keeping it as intact as possible. If your source is a worksite dumpster, a local dump, or a third party such as an individual who has just completed demolishing his own garage, you'll

An easy way to check what type of wood has been painted over when reclaiming materials from a structure is to pry up a sample and look at the ends or back.

SALVAGE WISDOM: OLD WOOD GRADES

Most companies grade their reclaimed lumber. Unfortunately, grading is not standardized, even informally. However, most resellers and salvage firms break their available stock down into three general grades: premium high end; mid-grade; and lower quality. These grades, or rating levels, reflect the number of visual and structural imperfections. High-end wood will feature even graining, with few if any knotholes or obvious imperfections. The color will be uniform. Mid-range lumber will have less consistent grain patterns, and the appearance will feature some telltale variations in color. Boards rated on the low end of the scale may have significant grain distortions and visible imperfections, and the color throughout may vary wildly, even over the span of a given board.

Don't get locked into thinking that grade determines the board you should buy. Although grading will reveal, to one extent or another, the structural integrity of the lumber, even low-grade boards will hold up fine as wall paneling or flooring. As far as appearances go, it's simply not as easy as good, better, and best.

What the appearance of any given wood stock means for you depends greatly on where and how you plan on using the wood. For instance, if you are installing reclaimed flooring or paneling in a country-style home to complement a rustic look, knotty pine or wormy chestnut may be ideal—even though they would be graded at the low end of the scale because of all their imperfections.

Sand back a spot of finish when you evaluate wood you are considering buying.

probably need to clean and prepare the wood before you can safely and efficiently use it.

The number-one risk in using reclaimed wood is hidden metal in the lumber. Striking a sunken screw when sawing through a reclaimed 4 × 4 for a trestle table is not only disconcerting, it's also very dangerous. The other important consideration is appearance. One of the main reasons anyone chooses to use reclaimed wood is because older wood is often more attractive. Depending on what you're using the wood for, you'll have to decide if you're going to sand down the surface completely, clean the surface, and stabilize the existing finish under a coat of clear sealant, or simply leave the finish as-is. A lot depends on where you got the wood from and what shape it's in. If you've rescued exterior redwood siding grayed from years of exposure and want to use it as interior wall paneling, you'll probably completely clean and seal the surface. On the other hand, reusing plank floors from a tobacco barn as new flooring in your great room may mean keeping that patina of age that was built up over decades of smoke from a curing fire. Other preparation may include cutting or ripping lumber to size for your purposes, or otherwise sawing to the dimensions and lengths you need.

But no matter what you do to the wood, it's wise to exploit as much of the innate beauty of the material as possible. If you're going to the extra effort to use reclaimed wood, you might as well use it in a way that best showcases the unique aspects of older lumber.

RECLAIMING WOOD TRIM & SIDING

The siding on older buildings is one of the best sources for reclaimed wood. The planks are usually long and plentiful enough to supply all the wood you'll need for just about any purpose you can dream up. If the siding was correctly installed, chances are that the lumber is structurally in good shape (although you may be faced with the occasional warped or cracked plank).

The process of reclaiming wood siding is a fairly simple one, especially if the interior walls are not clad. But even if they are, you simply remove the interior walls with a hammer and pry bar, as well as any insulation lining the wall cavities (do not handle asbestos or urea formaldehyde insulation yourself—they require professional abatement and disposal). Once you have access to the inside of the siding, removing it is just a matter of some careful hammering and prying.

Start by rapping the topmost board as close as possible to where it attaches to a stud. Hit only as hard as necessary to work the nails out, but not hard enough to crack the wood. Hammer until the siding board releases from the stud. Move outside and deliver a quick, sharp blow at the nail location, to slam the siding back down against the stud and free the nails. If you can't easily hammer the siding away from the stud without cracking the wood, use a pry bar to work the plank away from the framing. Remove nails as you work and use a helper to hold up the plank once you get to the last set of nails.

When you remove a board, check it for any nails that might have been left in it. Stack the siding in a convenient location, using bolsters every few feet of length to separate layers and protect against damage.

Trim pieces are even easier to remove, especially if they've been in place for a long time. Normal settling tends to loosen trim fasteners over long periods. That means you won't need to use a lot of force to remove the pieces, and you could damage them if you do. For heftier trim, you can use a simple crowbar to pry the piece away from the wall at nail points. For more delicate pieces, such as detailed wood crown or dentil molding, you can tape or otherwise pad the end of the pry bar so that it doesn't mark or crack the molding.

In any case, remove siding or trim to a heated indoor location. Storing these pieces—usually only painted on one side—in an outdoor location or where humidity and temperature vary widely may cause warping or damage to the unfinished face.

A pry bar can be a big help in removing siding that is held by stubborn nails. Use steady, even force rather than quick and hard levering. Exercise patience and the siding will eventually come free without cracking or other damage.

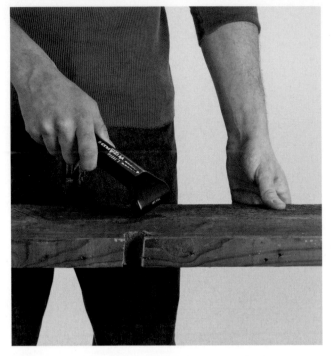

Finding the nail is the first step to removing it. That's why a handheld metal detector is a must for checking reclaimed lumber. A hidden nail or screw can destroy a saw blade and cause serious injury to the operator.

Nippers are a good choice for removing nails carefully, to limit damage to the board face. However, they will not work on sunken nails or flush nail heads because the pinchers need room to close over the nail body to pull it out.

Extractor pulling pliers are specifically designed for maximum leverage. They are quite effective for removing nails and other fasteners with a minimum of damage to the face of the wood. They are also easy and quick to use, but less effective on stubborn nails in hardwood.

Use a cat's paw for large, stubborn, embedded nails. Tap the slot of the tool under the nail head by rapping on it with a hammer. Pull back on the handle. Use a scrap of wood under the cat's paw head for extra leverage. A cat's paw is a blunt instrument and a last recourse, usually reserved for rough or hand-hewn timbers because it will likely mar the surface.

The time-tested method of prying nails out with a claw hammer works as effectively with reclaimed wood as it does with new lumber. Use a scrap of plywood or other soft pad under the head of the hammer to increase leverage and protect the face of the wood.

SEPARATING LAMINATED LUMBER

Separating laminated members is often necessary to salvage pieces for reuse. For instance, delaminating a header of sandwiched 2 x 8s can yield all the usable wood you'll need for a table. Pry laminated pieces apart with long pry bars or crowbars, by simultaneously pushing and pulling to create separation. Separating larger laminated sections will be easier with two people, providing extra muscle to force the pry bars in opposite directions.

SALVAGE WISDOM: MANAGING MOISTURE CONTENT

Moisture is an issue with reclaimed wood just as it is with any other type of lumber. Luckily, most reclaimed wood has had a long time to dry out. Unless the wood has been left outside in extremely wet conditions, it should be fine for immediate use. If you suspect that it has been overexposed to moisture—especially if you're salvaging the wood yourself—it's wise to check the wood with a moisture meter before using it in a final installation. In most cases, you want the wood to register under 10 percent moisture, but the drier the better. It's also smart to acclimate reclaimed wood in the final space before installing it as a matter of course. Leave it in the space for several days to a week before using it, to help the wood adjust to the exact moisture levels and temperature within that space.

RECLAIMED WOOD FINISHES

One of the key decisions you'll need to make when reclaiming wood of any type—from flooring to doors to cabinetry—is whether you'll preserve the existing finish or refinish the wood to one degree or another. It's a decision made more complicated by the fact that old wood, like other salvaged material, often bears a handsome patina of age. Your decision to refinish or leave as is will depend on how handsome that patina is, how stable the surface is, and how well the look might integrate into your home.

The first step is to determine what type of finish was used on the wood. If the wood was reclaimed from an old rural building, such as a barn, the wood may not have been finished at all. This is especially true of exterior elements such as siding planks. In most other cases, though, you're likely to encounter one of several common finishes.

PAINT

Salvaged casework, cabinetry, wainscoting, and other wood details were often painted. Painted vintage surfaces are not likely to be in good shape. But that doesn't mean they need to be stripped entirely. Painted vintage wood is most often repainted, because if the original coat of finish was paint, it quite possibly seeped deep into the wood's pores. Consequently, stripping it entirely may require removing a significant layer of wood and may be extremely difficult if the wood you're working with is detailed, such as crown molding or period casework. Perhaps the easiest way to deal with painted reclaimed wood is to opt for a distressed look. Simply removing loose paint and sanding down to the wood or primer in spots creates a desirable weathered look that can be stabilized by a coat of spray-on polyurethane sealer. Keep in mind that older paint may harbor lead, and you should always test painted reclaimed wood for lead—right down to the bottommost layers.

PENETRATING OIL

Oil finishes, such as tongue or linseed oil, are a little difficult to detect. Unless they've been polished, the appearance will be flat and natural. Depending on how old the wood is, the oil may have long since dried out of the wood. But if an oil finish still exists, it can be removed with mineral spirits. However, if the wood was stained before being oiled, the color will have to be deep-sanded out if you want a different color in the final appearance.

WAX

Wax has long been a finish of choice for furniture makers, but you may find it on casework, cabinets, and some flooring. The finish may be matte or carry a glossy sheen, depending on whether the waxed surface was polished or not. You can check if the finish is wax by rubbing with fine steel wool in an inconspicuous area; if the steel wool clogs with a gray smudge, the finish is wax. If the wax finish is still completely covering the wood (wax erodes over time), you can leave the finish in its existing state. Surfaces on which a wax finish has been worn away can be refinished with a new application of wax. However, if you prefer a whole new look for the wood, you'll need to remove any wax with mineral spirits.

SHELLAC

Used widely before 1930, and less often thereafter, shellac has a distinctive appearance, especially as it ages. You may find woods with shellac finishes intact, featuring a smooth surface and deep, rich amber undertones. But more likely, you'll find a pitted surface that often resembles an orange peel texture and that is tacky to the touch. Shellacked surfaces that have not deteriorated are often left as-is to take advantage of the beautiful look. But shellac that has degraded is usually removed.

A distinctive puckered and dimpled surface called orange peel *is the telltale sign of aged shellac.*

VARNISH AND POLYURETHANE

Varnish is an older style of finish that was used most commonly on furniture, although it has also been used on cabinetry. Polyurethane is a more modern finish. Both are hard coatings and usually yellow with age. They can cover stains or the natural wood. Pure old varnish (types that don't contain polyurethane) can be recoated with a new application of varnish. A degraded polyurethane surface must be stripped and replaced.

REMOVING SALVAGED WOOD FINISHES

Where you are certain of the surface coating on the wood, you can simply use the appropriate method for removing the material. Where the finish is in doubt, or where you just want to get a sense of what the finish actually is, use a hierarchy of potential stripping agents starting with the most benign and working up to the strongest.

Start by simply cleaning the surface with a dilute mixture of warm water and dishwashing soap to remove superficial dirt. Dry the surface thoroughly before assessing the condition of the remaining finish. Remove more stubborn dirt, grease, and wax with an application of mineral oil. Wipe the surface of the wood with a clean cloth dampened with the mineral oil, or use a mop to clean floors. Denatured alcohol will remove undesirable shellac finishes and wax, and is applied with a cloth in the same way as you would mineral oil.

Use acetone or lacquer thinner to remove lacquer. If you're sure the surface is lacquer, buy a thinner specially formulated for lacquer removal. You can remove varnish and polyurethane with special chemical strippers. However, these products are toxic formulations that should be handled with extreme care and safety precautions, including using a respirator, eye protection, rubber gloves, and appropriate work clothes. Follow the manufacturer's instructions for use to the letter, and thoroughly ventilate any space in which you're using these products. In most cases, however, it's just as easy and less dangerous to strip varnish or polyurethane by sanding.

Clean it before you strip it. A simple mopping with mineral oil can reveal the true nature of a wood floor.

STRIPPING PAINT

Aged paint can be a challenge to remove from wood surfaces. Fortunately, there are many different ways to strip paint off wood. But before you try any of them, test a vintage painted surface for lead. If the test comes up positive, remove the paint with a stripper formulated for use on lead paint. These usually soften and bond the paint to a fabric or paper strip so that the paint can be removed and disposed of without flaking or creating dust.

HEAT STRIPPING
A heat gun can be a quick way to remove paint from flat wood surfaces. Move the gun along the surface, heating the paint just enough to soften it and make it release from the wood. Follow along with a putty knife as wide

Heat stripping is quick and effective on paint, but keep the gun moving to avoid damaging the wood.

as practical for the wood surface. Using a heat gun takes a bit of practice, because if the paint isn't hard enough, you'll have to put a lot of effort into scraping with the putty knife and you risk gouging the wood. Too much heat, though, can make the paint overly sticky and gummy, making it difficult to gather and remove from the knife. A heat gun should never be used on paint containing lead, because the heated paint will release lead fumes.

CHEMICAL STRIPPING

There are many different types of chemical strippers, but most are toxic. Newer types, including soy-based formulas, are less so, but safer strippers tend to work more slowly and are somewhat less effective for stubborn paint. In any case, check labels for required safety precautions and ventilation requirements. The strippers are either paste or liquids, although paste strippers are easier to handle and apply. They are applied following the manufacturer's directions, but most are brushed on, allowed to sit, and then scraped off with a putty knife or other scraping tool. Wood covered in several layers of paint often requires more than one application of stripper. Using a chemical paint stripper is a messy job, so be sure to set up an adequate work area.

SCRAPING AND SANDING

Sometimes the quickest and most efficient method of removing paint—not to mention other surface finishes—is by sanding and scraping it off. This is especially true on flat, unadorned surfaces. Painted wood surfaces are usually scraped for repainting rather than to remove all the paint. Scrapers come in a variety of shapes and sizes, from wide flat types to tiny V-shaped scrapers for cleaning out grooves and other woodworking details. Sanding a painted surface is a more complete option that usually entails removing everything down to bare wood. A properly sanded wood surface can be repainted, stained, or refinished natural. Light sanding may be all that is necessary to prepare a surface for a new coat of paint. As with scrapers, there are specialized sanding tools for the nooks and crannies of casework, cabinetry, furniture, or molding. These include sanding cords that are like twine made of sandpaper and oval sanding blocks for getting inside curved wood pieces.

Chemical paint strippers can be effective, but they need to be applied liberally for best results. If only one or two coats of paint cover the wood, you may be able to remove the paint down to the wood surface in one pass. Where the wood has been painted in many layers, you'll most likely need more than one application.

Specialized scrapers are ideal for removing loose paint and quickly preparing a simple reclaimed wood structure—such as a wood window frame or vintage casework—for repainting.

Checking reclaimed lumber for rot; cutting rot or other damage on the end of a reclaimed board is the most common repair you'll need to make. Mint Images/Tim Pannell/Getty Images

RECLAIMED WOOD REPAIRS

When it comes to using reclaimed wood, whether it's flooring, an antique panel door, cabinetry, or some other type, there is a fine line between interesting and attractive surface imperfections and ugly, structurally compromising damage. Getting the most out of salvaged wood means being able to assess and fix any problems that fall on the wrong side of that line.

ROT

The wood reclaimed from older buildings may have been exposed to the elements or adverse conditions for a very long time. That's why inspecting any vintage wood for rot is a key part of the reuse process. Probe suspect sections with an awl to determine if rot has established within the wood. Often, though, rotted wood sections will be quite apparent to the naked eye. In some cases, such as antique paneling or casework, rehabilitating the piece will entail cleaning out the rot and sanding or chiseling down to stable wood. Working with a flat surface such as plank flooring or siding, crosscut the surface back from the rotted edge.

Certain situations, such as salvaged casework with rot at the bottom or a plank that has rotted along the edge near the center of its length, may call for a more creative solution. For these challenges, a number of wood consolidation products are on the market—specifically two-part epoxies, mixed together to form an infiltrating liquid that is poured or brushed over the rotted surface. The liquid penetrates the rot and dries to form a solid structure that can be sanded or formed with a surform tool, file, sandpaper, or other implement. If the rot has eroded sections of the wood, the consolidation mixture can be used as primer for an adhesive wood putty. The putty is also a two-part formulation—a resin paste that is mixed with a hardener. The putty is applied with a putty knife to replicate the missing section of wood, and then allowed to dry. When it's dry, it is sanded to match the profile of the surrounding wood. The wood is usually painted after epoxy consolidators or wood putty have been used, unless the rotted section was small or will be hidden from view—in which case the wood can be stained or finished natural.

DENTS

Dents in a wood surface are not charming indications of age like some scars or wormholes. Fortunately, you can easily remove dents, even in a finished wood surface. Moisten a clean, white cloth with distilled water, and place it over the dent. Lightly touch the dent area with the tip of a hot iron and the dent should come out. Deeper dents need to be filled with wood putty. Use tintable wood putty and tint it to a shade to match the wood species or existing finish. If the dent is in a noticeable area, you may want to spend some time with different color tints and a small paintbrush, and re-create the grain pattern over the dent.

GOUGES, NAIL, AND SCREW HOLES

Whether you leave or repair scratches or holes in a wood surface is largely a matter of the look you're after. If you want a distressed, antique appearance, chances are you'll leave many of these small imperfections. However, if you're using the wood specifically for the grain and appearance of the species, you'll probably want to repair many, if not all, of the modest blemishes in the surface. Use sandable wood filler to fix these surface flaws. You can purchase tintable fillers that will allow you to add color and help the substance blend into the surface, or choose a pre-tinted filler. They are available in a range of shades to match different wood tones. Press the filler into the gouge or hole, level it off, and allow it dry. Then lightly sand the surface to complete the repair.

Nail holes, gouges, dents, and scratches often can be fixed with just a little bit of wood filler.

SALVAGE WISDOM: FOOTWEAR FOR SALVAGED DOORS

Rot, especially rot along an unfinished bottom edge, is a common problem with antique doors. Left untreated, the rot can spread. Solve the problem by removing the rotted portion and cladding the bottom of the door in a protective metal "shoe." Mark the cut line right above the rot, snapping a chalk line across the door face. Use a straightedge clamped to the door and positioned so that the circular saw blade cuts along the cut line. Sand and seal the cut edge, and cut the shoe—available at home centers and hardware stores—to the width of the door. Screw the shoe onto the bottom edge and install the door in its opening.

RECLAIMING WOOD FLOORING

A home's flooring is all about beauty and comfort underfoot. No flooring is more beautiful and comfortable than wood, and no wood offers a more unique character than reclaimed wood.

Fortunately, old buildings often yield wonderful wood flooring. You can find reclaimed strip, plank, and historical pegged floors, and the tile alternative: end-grain flooring. Siding and paneling from older buildings can also be repurposed into new wood flooring.

No matter which you choose, you'll either reclaim the material yourself or buy it from a reseller. If you have the opportunity, wood flooring is some of the easiest material to salvage from a building. Start by clearing the room and removing baseboard moldings. Remove the first and second row of boards on the tongue side (this may mean destroying one or more boards). Use a pry bar on damaged flooring so that you have clear access to the rest of the flooring. Remove the remaining planks or strips by sliding the pry bar under the tongue next to the nail points, and slowly and carefully popping the strip or plank up.

If you're rescuing siding, it's relatively straightforward to mill it for use in a floor. It takes time and equipment, but doing the work yourself can save you a great deal of money. Of course, resellers are the easier route. These companies often remove entire floors and sell them in bundles. In other cases, the seller will group boards with the same measurements and similar appearances. They may even pre-finish the boards.

Installing a reclaimed wood floor is usually the same as installing a new wood floor. However, an unusually wide plank floor or pegged wood floor may require face drilling and plugging (see page 119). In the end, you'll choose to either keep the existing finish as-is, or sand and refinish.

Reclaimed barn siding was lightly sanded before being installed for the floor in this stunning bedroom. The color variations in the floor planks provide endless fascination.

RECLAIMING A VINTAGE WOOD FLOOR

Reclaiming some wood building materials directly from an existing structure can be a challenge. Structural members such as timbers and beams must be extracted carefully to prevent wholesale collapse. Others, such as siding, are arduous to remove and will need exceptional amounts of prep to be reused. But wood flooring is one of the easiest and most-rewarding materials to salvage. Do it right, and you may not even need to refinish the reclaimed flooring after you reinstall it.

The process is fairly straightforward and is basically the reverse of installing a wood floor. Start by clearing the room you're working in and remove shoe or other base moldings. Remove the first and possibly second row of boards on the tongue side of the floor. This may require destroying one or more boards to gain access. Do that by using a pry bar on damaged flooring, or a circular saw and pry bar, to cut into and tear out boards so that you have clear access for prying out the adjacent rows.

Removing planks from that point is relatively easy. Slide the tongue of a pry bar under the tongue of the plank next to a nail, and pry the nail up. Do this at each nail location until the plank is completely loose. Then pry up and toward you to release it from

Stack reclaimed flooring neatly, well supported by bolsters. If you are planning on using the flooring with its existing finish intact, separate the layers with building paper.

the next row. Continue removing planks, taking care not to damage the tongues as you remove the boards. Remove all the nails as you work, and check boards for nails before finally placing them neatly in stacks separated by bolsters.

RECLAIMING WOOD DOORS

Wood doors are the low-hanging fruit of reclaimed building materials. Easy to get at and just as easy to remove, these essential building elements can be plucked from just about any building or structure whose time has come to an end. More contemporary versions become available as homeowners tear down walls and reconfigure layouts and floor plans as part of major home renovations.

Whether you're interested in a set of distressed doors for your kitchen cabinets, looking for a sturdy entryway with distinctive style, or just trying to match the interior six-panel doors in your three-bedroom Victorian, you'll find a spectacular number of alternatives available through many different sources.

SALVAGE WISDOM: ACCLIMATING RECLAIMED FLOORS

Wood planks or strips that you reclaim should already be thoroughly dry, because they have likely been inside for decades. However, in some cases you will have rescued the wood from a dilapidated building so far gone that the floors were exposed to the elements for quite a while. In other cases, resellers may have stored reclaimed wood flooring outside. Wood flooring in these situations can absorb moisture. But even if it's just a matter of temperature change, you need to give the flooring a chance to adjust to its new environment. Store reclaimed wood flooring in the room where it will be installed for at least 24 hours prior to installation.

Salvaging doors right from the source is often the most desirable option because it takes very little time and effort (you won't need much beyond a measuring tape, hammer, pry bar, screwdriver, and half an hour). A far wider selection can be found at salvage companies. But no matter what source you use, expect to be faced with an almost-overwhelming number of variables. Door styles have changed frequently through time, and many different styles were used in any given historical period.

Beyond their obvious use as new or replacement units, doors can be recycled as pieces of furniture. A basic eight-panel front entryway unit can be refinished in your favorite color, set on top of two painted sawhorses—or better yet, legs made of reclaimed staircase newels—and covered with glass to make a beautiful desk. A panel door can be refinished and distressed to serve as a wonderful country-style headboard for a bed. They can even be taken apart for reuse. Older doors are usually solid wood, with stunning grain patterns and coloring that can bring a handsome appearance to plain bookshelves, benches, and other small projects.

But by far, the biggest reclaimed use for doors is as doors. Some fit the current standard interior door widths of 28, 30, or 32 inches, or the exterior standard of 36 inches. Others, such as barn doors, are built to unique measurements and either the opening or the door will have to be modified. Some doors come pre-hung—that is to say, nested in their own door frame—but most of the doors that are reclaimed are standalone units perfect for retrofitting into an existing doorway. How much work you need to do to cut down or otherwise customize the door for your location will be an important consideration in selecting a new door. Buy a door that needs to be drastically downsized to fit your opening and you risk making the door look odd. On the other hand, you should simply never buy a door that is too small for the opening. These points are true whether you're looking for interior, exterior, or cabinet doors.

WHICH HAND?

Obviously, it's essential that you know how the door you're considering will go into the opening. Any exterior or interior door with a handle is generally categorized as either left- or right-handed. The term specifies which side the handle is on, and which side the hinges are on. If you face the door as it opens toward you, the "hand" of the door is the side the doorknob is on. Some reclaimed doors can be flipped to fit in an existing opening if necessary, while others can only be installed one way. A door's "handedness" is more important than you might imagine, because where the handle is determines which side the hinges are on, which in turn determines how the door opens. Preferably, you don't want a door to open so that the person walking through it is sandwiched between a wall and the door, or is facing a wall. Doors should swing into the room you're entering or leaving. Front doors, for instance, generally open into the house.

Cabinet doors must be chosen with the same issues in mind. You need to make sure the hinges and handles are on the correct sides for the cabinets you're looking to retrofit, or that the hardware on the doors can be easily swapped without ruining the look of the doors or cabinets. Size will be another major concern if you're searching for cabinet doors, although you may be able to adapt larger doors to your purposes, if you change from face-frame or surface-mount hinges to concealed or inset hinges. Obviously, though, no matter what type of door you're looking for, a measuring tape is your first and most essential partner in any door salvage shopping trip.

RECLAIMED DOOR TYPES & STYLES

Not all doors are created equal. In fact, door styles are as varied as any architectural element. If you're adding a reclaimed interior door, it's often easiest to choose one with an identical or similar appearance to the other doors in the interior, although adding a set of French doors to a dining room or saloon doors to a kitchen can change the appearance of the room and the home. Reclaimed cabinet doors should generally all be the same. But using a reclaimed exterior door is your chance to add a focal point. Refine the look of the home with a stately walnut door containing a half-lite of beveled glass. Create a Tuscan look with an arched, board and batten door fitted with oversized iron hinges. The possibilities are nearly endless, but your decision begins with the types of doors available.

PANELED

This is a common style that has been used down through time for both exterior and interior openings. Different types and arrangements of panels indicate different periods in history, although most panel designs can be

SALVAGE WISDOM: COMMON DOOR PANEL STYLES

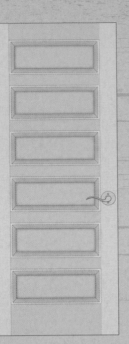

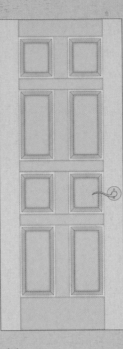

Horizontal Raised Panels

This look is strongly identified with the Prairie School and Craftsman design movements, and it is best suited for modern house styles. The doors usually feature four, five, or six horizontal panels stacked between the stiles. The look is subdued but pleasing.

Eight-Panel Colonial

Although not distinctively different from some Victorian and contemporary doors, the Colonial style was defined by alternating square and rectangular vertical panels. The look is adaptable and is used in many different home styles.

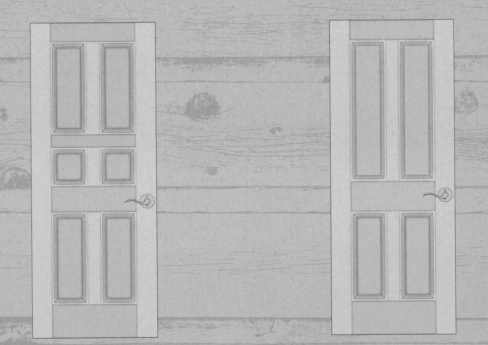

Six-Panel Victorian

The configuration of Victorian panel doors differed from earlier and later versions in subtle but noticeable ways. Panels were often stacked at irregular distances, as this example shows. Victorian-era door panels were also much more likely to be arched.

Four-Panel Traditional

This configuration, or others similar to it, has been in use from the mid-twentieth century through to the present. It's a visually pleasing if uncomplicated arrangement of two panel pairs, one larger than the other.

This six-panel interior door is made of fir, but it blends well with the reclaimed oak casing.

Wood panel doors are becoming very rare for exterior locations because they are not as efficient as non-wood doors and are more prone to movement and warpage.

adapted to many different types of architecture. The doors are crafted of rail-and-stile frames, with panels that are either flush (flat), recessed, or raised. Flat-panel doors are the plainest and more commonly used as interior rather than exterior doors. Raised-panel doors offer the most ornamental appearance and come in a wide variety of panel configurations. The same is true of recessed-panel doors. But recessed panels appear somewhat less ornate and plainer than raised-panel doors do.

BOARD AND BATTEN

Think Old West, country, or rural. This is the most rustic style of door, with vertical boards forming the body of the door, and battens running horizontally and/or diagonally to hold the whole thing together. This is a distinctive style, appropriate for a limited number of home styles. The doors look odd when painted, so you have to accept the country appeal on its own merits. Arched versions are more "wine country" than "back country," but still would not look right on most contemporary, traditional, or modern

structures. Style aside, the insulating value is very low. However, the right board and batten door can add to the charm of a cottage or a summer home, and they make excellent doors for fences bordering flower gardens or other well-landscaped yards.

FRENCH DOORS

Dating from the seventeenth century, French doors were common enough throughout history that they now make regular appearances as part of salvage companies' inventories. The original idea behind the French door was to let light pass freely throughout the living space. Although they still serve that function today, you'd be wiser to use reclaimed French doors on the interior, because the single panes don't have the insulation value that a more solid door does. All French doors feature top-to-bottom glass, but the number and arrangement of panes create a remarkable assortment of styles. Typically, the more panes, the more sophisticated the look. But more panes are also harder to keep clean—a point worthy of consideration when choosing among reclaimed French doors. The doors are used both as single and double doors, although the most impressive use is as a set, for instance in the entryway to a formal dining room. A single French door with a grid of lites can also be turned into a rather stunning table or desk. Or mount one on the wall, filling each pane with a different picture, to make a visually captivating photographic montage.

PANEL LITES

Putting a window in an otherwise solid door has long been a way to let the sunlight into a home and take advantage of a view (or keep an eye out for visitors). Panel *lites*—the term used for panes of glass in doors and adjacent areas—come in as many shapes and sizes as solid-panel doors do. Exterior doors most often have glass in the top half of the door, and the lite is frequently made of art glass. Using beveled, machined, or stained glass in a door's half lite allows sunlight to penetrate, but obscures the view to maintain privacy. Other styles of glass, such as a fan lite at the top of the door, are indicative of certain styles and particular to certain tastes.

CABINET DOORS

New cabinet doors can give the whole kitchen a face-lift, and reclaimed cabinet doors come in as many different styles as their contemporary counterparts. They're also an inexpensive alternative to replacing or re-facing your existing cabinets. Cabinet doors come in flat, raised-panel, and recessed-panel profiles, as

well as versions with glass inserts either divided or undivided by muntins. Glass inserts can be frosted, beveled, machined, or plain glass, all of which can bring a distinctive look to the kitchen. Selecting the right salvaged cabinet doors can be a bit of a challenge. You have to find a size that will work, a style that suits your taste, and enough doors for all your cabinets. However, the cost will likely be a bargain, and you can often get great impact out of just replacing upper or lower cabinet doors.

BI-FOLD AND CLOSET DOORS

Upgrade a room's look with a nice set of vintage closet doors, or make long bookshelves out of disassembled solid-wood bi-fold doors. Louvered, solid, and glass folding doors are all commonly reclaimed from older homes and newer remodels. Stripping and staining solid-wood closet doors can be a great way to add some style to a bedroom, while painting louvered doors in vivid colors can be an excellent way to jazz up an otherwise boring area of any room.

POCKET DOORS

These space savers have been around since Victorian times, and they are still fashionable and ideal for tight rooms. Salvaged pocket doors include those that look like French doors, simple solid versions, and panel doors. It's possible that you may be able to convert a pocket door into a hinged door, although they are often too short for standard openings. If you plan on reusing a pocket door you've reclaimed, make sure the tracks are in good shape. Track quality has varied over time, and track failure can lead to a lot of frustration. If you have doubts about the tracks that come with the door, look to replace them with newer hardware.

WORKING WITH RECLAIMED TIMBERS & BEAMS

Structural timbers and beams are some of the most dramatic examples of reclaimed wood. Timbers are steeped in history because reclaimed structural members are usually some of the oldest wood you'll find. Big and substantial, they have an impressive physical presence. Their impact is made even more powerful by exceptional variations in surface texture. Evidence of hand crafting is one of the key allures to using reclaimed wood, and few pieces bear the marks of a craftsman's tools and expertise as blatantly as salvaged timbers and beams do. Many were hand hewn with axe, adze, and hand plane, creating a rough and unique surface full of character.

But even beams and posts that were milled with industrial saws retain arresting surface textures. Because most served purely structural and functional roles, it was considered a waste of time to plane or sand down to a perfectly smooth surface. Thick beams, timbers, and posts consequently often have a rough-sawn surface, or may exhibit the scars of circle or skip sawing. Each piece has a vivid historical tale written on its surface.

As charming as they are, though, these rough, crude surfaces tend to collect dirt and grime, especially in exposed locations and over the long periods of service they've commonly seen. Given all the collected

A reclaimed thick eucalyptus plank is a more appropriate mantel for this modern fireplace than a rough-cut full-scale timber would have been.

grunge, using one of these members in an interior design project usually means thoroughly cleaning it first. Once you get down to a stable wood surface, you'll need to make a decision about what kind of appearance you ultimately want. Most people look to strike a balance between a timber surface that is pleasant to the touch and the eye, and one that retains the wood's original character, beauty, and woodworking marks. Depending on how the timber was manufactured in the first place, and how hard the years have been on it, reviving the wood for reuse in home design can range from a simple cleaning and light sanding to a full-scale planing and smoothing.

How fully you finish the surface depends in no small part on what you have in mind for the piece. Because vintage timbers and beams are so sturdy and inherently sound, they are often incorporated into homes as exposed beams or support columns—either as purely decorative elements or structural members. But there are many other uses for these unique pieces. Timbers, posts, or beams of any sort make show-stopping fireplace mantels, especially when fronting a stone fireplace surround. They also work well over more contemporary steel or brick-fronted fireplaces as a contrasting design element that calls attention to its own imperfections. In any case, the timber can be centered over the fireplace, or a longer member can be used to create an asymmetrical look as attractive as it is unusual.

Timbers and beams can serve a home design in many other ways as well. Sturdy headers can be turned into wonderful wall-mounted shelves. Barn or warehouse posts can be ganged to create a fascinating room divider, or split to make unusual handrails around a deck. And the wood can be reused in a functional way by cutting and fitting it to build a trestle table or countertop. But no matter how you reuse the piece, a salvaged timber is inevitably home-design gold.

The interesting character of many reclaimed timbers lies in the marks of hand-hewn surfaces.

A power washer is the quickest way to remove decades of grime and reveal the true nature of the wood below. Use a medium nozzle and pressure to quickly remove dirt and grime without damaging the wood surface. Hold the nozzle at an angle to the wood face, about 6" to 8" from the surface, and keep the spray moving.

If you want a very smooth and level surface, you may need to plane the timber. Adjust the blade on the hand planer with the adjusting knob, starting out with a shallow setting. Be sure to check for nails or screws first. Press down slightly on the front at the start of the pass, and on the back at the end, using a smooth, even stroke. Check the shavings to ensure you are taking off very little wood in an even layer.

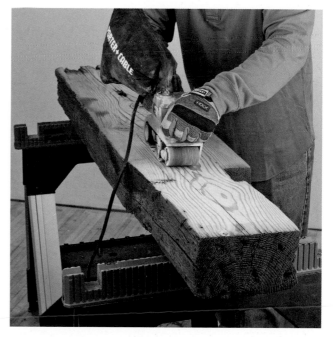

A power planer can make quick work of an extremely rough timber surface. Start with the blades adjusted to the shallowest setting. Set the planer's front shoe on the surface, start the planer, and let it get up to full speed. Then guide the planer forward. Maintain easy pressure and control the planer to keep it level at the end of the stroke and to avoid taking a chunk out of an edge.

For the smoothest surface possible, you can sand the timber. Even with a belt or palm sander, this is a tremendous amount of work. Sanding is usually only required for specialized finishes or if the timber will be used as a high-exposure surface in the home.

REVIVING VINTAGE CABINETRY

Old cabinets abound because storage has always been a premium. Many older units feature wonderful features such as tip-out bins and spice-rack drawers. And they are often constructed with details that last—like solid wood panels and dovetailed joints. However, be aware that reclaimed cabinets are best used as accent storage; finding a suite of correctly sized old cabinets for an entire kitchen would be highly unlikely.

The good news is that these cabinets can be bargains. The original owner has already accounted for the cost of craftsmanship and materials. Salvage companies are often willing to negotiate to a very reasonable price because older reclaimed cabinets aren't easily adaptable to existing spaces.

Chances are that you'll have to do a bit of sprucing up to any older cabinetry you reclaim. This can range from a thorough cleaning to wholesale structural reinforcement or rebuilding. Start with the finish. Sand or strip the cabinet down to bare wood if you want a fresh look, or simply remove loose paint and debris and spray with a polyurethane sealer for a distressed look.

Before you work on the finish of the cabinet, however, you should thoroughly inspect the structure

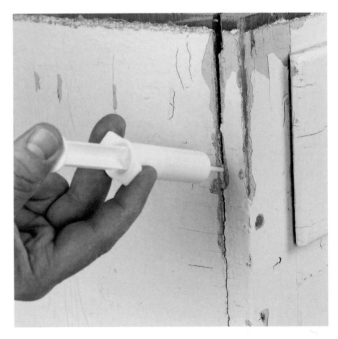

Brace the corners. Chisel a mortise in the top for a metal L brace and screw the brace down with ½" screws. Alternatively, cut triangular glue blocks out of square pieces of 2x stock. Check the cabinet for square, clamp the box, and glue and screw the blocks into each corner.

Reconstitute the joints. Box or butt-joints can separate over time. Restore a separated joint by injecting wood glue into the joint with a glue syringe. Clamp the joint until the glue dries. You can also use this method for drawers.

to identify any structural problems. This will entail removing doors, drawers, and all hardware.

The best cabinets are made of dovetailed joints, and those are likely to still be in good shape. Box or butt-joints are more common but less secure, and chances are if the cabinet is fairly old, you'll need to reinforce the joints to one degree or another. You may also need to square up the cabinet so that it doesn't sag or hang out of true.

RECLAIMING & REUSING PALLETS

At first glance, wood pallets might not seem like the ideal target for salvage and reuse. Functional and clunky, they don't lend themselves to an obvious use beyond the purpose for which they were originally intended. But first glances can be deceiving; pallets are excellent candidates for upcycling, used whole or broken down.

The size and shape of most pallets makes them ideal to craft similar-shaped furniture, such as large coffee tables or beds. They're also ideal for exterior structures, such as the frame of a garbage can enclosure or compost bin, where the slatted structure will allow for essential airflow. More commonly, pallets are disassembled so that the raw wood can be reused in everything from wall paneling to raised beds.

You don't have to look far for pallets to reclaim, because they're everywhere. The US Forest Service estimates that almost two billion pallets are in use in America. In fact, pallet manufacturing accounts for the largest hardwood lumber use in America. Businesses small and large use this ubiquitous platform to ship just about everything that can be shipped. But pallets aren't long-lived. They are subject to excess weight, near-constant movement, and extremely rough handling. It

should come as no surprise that millions end up in landfills every year. That translates to a mindboggling number of pallets available to reclaim as your own. It's just a matter of finding and collecting them.

Although the obvious places to look are industrial areas with plenty of warehouses and manufacturing companies, larger companies may actually not be the best sources for pallets. It's possible that the pallets were used to ship chemicals and may have been subjected to chemical leakage and spills. For the same reason, supermarket pallets should be carefully checked for any food stains that could be evidence of bacterial contamination. You'll likely have more luck with pallets from small hardware stores, motorcycle shops, and other equipment retailers that receive new equipment on pallets.

Regardless of where you find the pallet, make sure that you follow the basic rules of upcycling. Don't go onto private property without permission, and don't assume stacked pallets are meant for discard. It's always better to ask if you can take pallets, even if they're in a dumpster (some companies maintain an agreement with recyclers who collect and recycle pallets from dumpsters).

Once you've collected the pallets for your project, it's a good idea to clean them up. You can use a power washer, or simply scrub the wood with a detergent solution (you don't know where the wood has been). Allow it to dry completely before using the pallet in a project or deconstructing it for the wood. A good sanding can make the wood look nice, but many upcyclers use the wood as-is, for its rough industrial appearance. The wood also takes paint well.

More than 1.8 billion wood pallets are used every day, and more than 450 million new wood pallets are made each year. Many will end their lives in a waste pile like this, ready to be spruced up and reused in many different forms.

➕ SAFETY FIRST

When reclaiming wood pallets, especially if you don't know where they've been, it's essential to closely inspect them. Check for obvious signs of contamination, such as oil stains or a chemical smell. It's also wise to look for stamps or other marks that might tell you what the pallet carried—which in turn tells you the potential contaminants to which the wood might have been exposed. Unmarked pallets were most likely used for domestic transport and are usually safe for reuse unless there are detectable signs (stains or smells) of contamination. Here are some of the markings you may find on pallets.

1. **IPPC** (International Plant Protection Convention). The IPPC certification symbol indicates that the pallet meets the specifications for international shipping. The symbol is usually accompanied by a two-letter code indicating the pallet's point of origin (such as "US" for United States or "CA" for Canada). The logo may be accompanied by a two-digit number, representing a certification number that allows the pallet to be traced back to the auditing agency. There may also be a four-digit number

that ties the pallet to the manufacturer and/or a company that treated the wood. Either of these numbers is a way to trace the origin of the pallet online. The wood in any IPPC-stamped pallet is pest and disease free.

2. **MB**. This stamp indicates that methyl bromide was used to fumigate the pallet (its use was banned in 2010). Avoid chemically treated pallets for home and garden projects.

3. **HT**. This stands for "Heat Treatment" and verifies that the pallet has been subjected to heating so that the core temperature was at least 132.8°F for at least 30 minutes. Pallets are heat treated to destroy insects and microorganisms.

4. **DB**. The initials stand for debarked, indicating the type of wood that was used in the pallet. It's a way to ensure that the wood does not carry invasive insect species or plant diseases. The IPPC requires all international pallets be debarked.

Avoid colored pallets. The color is often a signifier that the pallet was used by a pool supplies manufacturer or retailer, and that the pallet carried chemicals.

Above: *It's fairly simple to turn pallets into a queen-size bed, like the one shown here. The headboard has also been crafted of deconstructed pallet lumber.*

Right: *Pallet wood is ideally suited for use as outdoor furniture. Sand and paint it, like the outdoor sofas and tables shown here, and you can have a beautiful addition to your patio for next to nothing.*

CHAPTER 3

SALVAGING OLD METAL

An amazing diversity of metals can be salvaged from old buildings and antique structures. Usually, the older the structure, the more interesting and rare the metals you'll find. It might be a finely detailed tin ceiling from the kitchen of a 1900s house, or the heavy rail and fixtures on which an old barn door slides open and closed. Although many of these metal pieces are not the most obvious choices for reuse, they represent terrific options for some imaginative upcycling. Repurposed metal siding or fixtures can bring undeniable charm and unique design accents to an otherwise nondescript home. Even the more pedestrian metal plumbing pipe or sheet-metal flashing are candidates for reuses, such as in pot hangers and backsplashes. As with wood, the patina many old metal surfaces feature can be as appealing as a brand-new finish. The look can also be varied because most upcycled metal can be combined with salvaged wood, reclaimed stone, and other materials to create intriguing combinations.

Reclaimed metal comes in two basic forms: fixtures and accents that are often reused for the same role they filled originally, and raw materials—such as metal pipes, flashing, and siding—which will be repurposed as something entirely different. Most of the metal salvaged from older buildings falls in the second category. But that doesn't make it any less useful.

In the normal course of business, demolition companies and contractors sell most of the metal recovered from a building to a scrap dealer. The scrap dealer recycles the material, or passes it along to a third party—such as a smelting operation—for recycling. Traditionally, this has been one of the "green" parts of building demolition and renovation.

However, reclaiming and reusing these materials is even greener. When you repurpose metal grates, radiators, steel doors, and other similar leftovers, you are saving the energy it would take to melt them down for recycling. You're also saving the waste products inherent in the metal recycling process. That's one of the things that make reusing reclaimed materials in home design so attractive. The other, of course, is the actual appearance of what you create.

Reclaimed metal materials that are inherently attractive are the easiest to adapt into a home's design.

For instance, finding a new purpose for tin ceilings is not a huge challenge to the imagination. The same is true of cast-iron grillwork used over heating vents, as window security, and in other forms. Even radiators are easily pressed into service as eye-catching table bases because their forms are simply striking.

On the other hand, if you're willing to stretch your creative muscles a little, you'll find plenty of uses for more mundane salvaged metals. Plumbing pipe, whether it's copper, cast iron, black iron, or galvanized steel, can be reused in a number of innovative ways. And working with these plumbing fundamentals is easy because the skills you'll need are a snap to master, and the tools for working with pipe are probably already in your toolbox.

Whether you've picked out some rare copper tiles to grace a breakfast bar skirting, or have decided to create a set of bookshelves from a jumble of galvanized iron pipes, you'll soon discover another wonderful thing about repurposing metal building materials: they are relatively inexpensive. Because bringing these types of materials back to life involves both imagination and elbow grease, building owners and salvage firms alike are usually willing to part with them for a song.

Pressed sheet metal can be used for decorative projects, either as-is or in revived form.

A little prep work and painting is all that's needed to make these radiators an eye-catching base for a low table.

+ SAFETY FIRST

Reclaiming metal building materials such as gas or water pipes and tin panels can be rewarding because they are usually valued less by a property owner than any beams, timbers, and other high-profile pieces. But if you get the chance to salvage these materials, practice safe deconstruction procedures. Make sure the main gas valve has been shut off before working on gas pipes, and safely bleed the pipes into open air before removing them. Water lines should be drained before you disassemble plumbing, and avoid waste stacks and other plumbing that has anything to do with sewage or waste water.

TYPES OF RECLAIMED METALS

Metal	Profile	Appearance	Best Uses
Tin 	Most engaging form is tin ceiling panels. Older corrugated tin panels used on barn and water tower roofs make wonderful rustic or industrial wall surfaces.	Reclaimed tin ceiling patterns are unique and ornate; can be left distressed, stripped and painted, or left natural and sealed. Rare versions coated in brass or copper are showpieces. Tin roof panels are a rougher, raw look, often left distressed for a style statement.	Reuse as ceiling cladding, wainscoting, backsplashes, accent wall paneling, or interesting cabinet panel inserts.
Copper 	Most reclaimed copper is in the form of plumbing pipes, but you may find hammered copper tiles.	One of the most purely beautiful metals; can be sealed to retain its new, shiny red appearance, or left unsealed to collect an eye-catching matte jade-green patina over time.	Copper sheet and tiles make stunning backsplashes and wall accents, and copper pipes and fittings can be configured into shelving units, pot racks, and more.

Metal	Profile	Appearance	Best Uses
Iron	One of the most abundant reclaimed metals, available as cast iron or wrought iron. (Cast is poured into molds to make objects requiring precision; wrought is handworked into unique shapes.)	The forms of iron—from intricate strap hinges to radiators—can be elegant and unusual. The metal can be sealed and used bare or—the far more common method—painted.	Used as decorative accents, while sturdier brackets, pipes, fittings, and radiators can be used as structural support for shelving, tables, desks, and more.
Stainless Steel & Aluminum	The least commonly available salvaged metals because they are some of the most valuable—often going right from deconstruction to a recycler.	Both stainless steel and aluminum resist rot and corrosion, so their reclaimed appearances are often the same as new.	Sheets or panels of either of these can make stunning and durable tabletops, wainscoting, or backsplashes. Fixtures such as stainless-steel sinks can be reused as-is, and pipes and connectors can reused as table legs and other bits and pieces.

RECLAIMED TIN CEILINGS

Beautifully ornate tin ceilings first saw widespread use in the late 1800s. They're now used for purely decorative purposes, but they originally had both decorative and functional roles. They were handsome imitations of decorative ceiling plasterwork common in homes of the wealthy. They captured much of the intriguing detail of a plasterwork ceiling, with intricate stamped designs and surfaces that could include centerpiece medallions and showcase borders. But they were also fireproof surfaces in kitchens with suspect wiring, and where open flames were the standard for cooking and heating. The popularity of tin ceilings grew because the surfaces looked so handsome and ornate.

Tin ceilings through history have been crafted in hundreds of relief designs. That means salvaged ceiling panel sheets may have a completely unique look you can't find at retail today.

Reuse tin ceilings as a ceiling pan, or use sections to adorn any flat surface, from a half wall to a backsplash. These panels—or tiles—make wonderful wall coverings and are often repurposed as wainscoting. They can be used as kitchen cabinet inserts or even as countertops or table surfaces.

Most tin rescued from older buildings will not only be covered in degraded paint, but it will also likely carry a coating of grime from airborne pollutants such as the grease residue from cooking. At a bare minimum, the tin will require a good cleaning prior to reuse. More likely, it will need to be stripped down to bare metal and finished with paint or a sealant such as polyurethane.

As you clean and strip, you may be fortunate enough to discover a different surface than tin. Tin ceiling panels were actually tin-plated steel, to create a more structurally sound surface that was less prone to flexing. Less often, though, the panels were plated in copper or bronze. Both of these provided a highly desirable look, and they are usually covered in a clear finish and placed in a high-profile location to show off the beauty of the metal.

RECLAIMING A TIN CEILING

Tin ceiling panels are some of the easiest metal materials to reclaim from an existing structure. Although the work is dirty, it's not a challenge to your DIY skills and it's not even all that physically

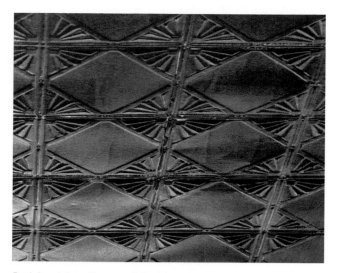

Reclaimed tin ceiling panels look best if you can retain some of the antique feel—otherwise, you're better off buying new material.

➕ SAFETY FIRST

Taking down a tin ceiling is not terribly hard work because gravity is on your side. But the force that brings the ceiling down can also bring down decades of accumulated debris. There are really two main safety hazards when deconstructing the ceiling: sharp edges and airborne material. Because tin ceilings predated the widespread use of asbestos as an insulating material, it's not a common concern when tearing out an original tin ceiling in a very old home. However, other material, including compacted dirt from the floors or attic above and mold, can create a lung-harming cloud. Always wear extremely tough work gloves and a thick, long-sleeved work shirt or jacket to avoid serious cuts from the edges of the tin panels. Wear a hat, respirator, and wraparound eye protection to ensure that the work doesn't take an unexpected toll on you.

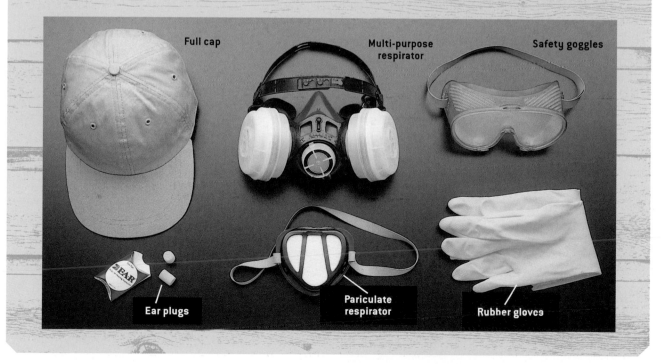

Full cap Multi-purpose respirator Safety goggles

Ear plugs Pariculate respirator Rubber gloves

demanding. Given the cost of reproduction tin panels, the few hours you'll spend taking down and reclaiming an existing ceiling will be time well spent.

Tin ceilings were originally mounted on a grid of wood lath; if you're reclaiming a ceiling in an older home, that's probably what you'll be dealing with. Fancier versions may also have cornice running around the seam between ceilings and walls, and may even feature flat molding, special profile pieces that separate a center square of more decorative field panels from an outer border of plainer filler panels. All of these pieces can be reclaimed and repurposed.

To take down the ceiling, start with the outermost pieces first, working your way toward the center. If there is a cornice, start with the corner cornice piece that isn't overlapped by any other. Pry it (this shouldn't take much effort because time and structural movement will have loosened the nails' grip) away from the wall, using a small pry bar with a thin, flat prying edge. Continue stripping pieces of the cornice, jimmying it at nail locations along the panels. Stack them on top of each other in short piles.

Next, you'll begin working on the filler panels if there are any, and the field panels if there aren't. Pry each panel down from the outside, taking care not to excessively bend the panel. As you pry, it will help if you pull the panel toward you. Once one or two edges are loosened, the panels should come down fairly easily. Any stubborn panels should come down with careful prying and pulling.

Watch for any unexpected debris as you take panels down off the ceiling. A lot of debris, dust, and insulation may come out as you remove panels. Work slowly and you're less likely to take a dirt shower.

Tin ceiling panels can be inset into a recessed area, such as this coffered porch ceiling, to turn them into a focal point.

Stack the panels in short piles and make separate piles for field and filler tiles. It's a good idea to secure each pile for transportation by binding it with twine or duct tape, after sandwiching the pile between top and bottom pieces of cardboard. Store the panels indoors, in a garage or shed, where they won't be exposed to moisture. Although it's likely that the face of the panels will be covered in paint, the backsides will probably be unfinished and susceptible to rust.

CLEANING & PREPPING TIN PANELS

Chances are that any tin ceiling panels you reclaim will need a good cleaning before they can be reused. In some cases, you'll be using the panels just as they are for the charm of a vintage appearance. In other situations, you may want to completely remove the old finish so that you can either display a copper or bronze surface to its best advantage, or repaint a tin panel.

There are two types of chemical strippers, those with methylene chloride and those without it. Non-methylene chloride strippers are much less toxic and can be used inside with modest ventilation.

Methylene chloride strippers, however, are very toxic and should not be used inside without appropriately rated ventilation.

Either way, the process of getting a tin panel clean is sequential—start with the mildest option, and work to the most caustic.

The first step, and the one all reclaimed tin panels, cornice, and molding should go through, is a basic washing. This will not only remove obvious dirt, but it will allow you to assess the integrity of the finish underneath that dirt. You may be surprised at how easily a little soap and water can bring a painted tin panel back to life. Use a mixture of lukewarm water and mild dish soap in a sprayer. Spray the panel and scrub lightly with a stiff nylon scrub brush. The dish soap will usually remove even stubborn layers of grease. Dry the panel immediately after you've washed it, using a clean, dry towel. Never leave a tin panel to air dry because that is just inviting rust.

A good cleaning will often reveal a still-attractive finish on a tin panel. But if you find loose or flaking paint, or simply hope to repaint the panel, you'll need to get down to a stable, rough surface to which the primer and topcoat can adhere. Use a drill equipped with a wire wheel to remove any loose paint, apparent dirt, and rough up the paint that is left. If you are planning on finishing the surface natural with a clear coat of polyurethane,

you should strip the paint using a chemical stripper or—the more environmentally friendly option—soak the panel in a hot water and soap bath. You'll need a large tub, but this solution can be incredibly effective for stripping paint off tin panels. You can polish the surface of a tin ceiling panel or border piece using a buffing wheel and polish specially formulated for whichever metal surface covers the panel.

A heat gun is usually the quickest way to strip multiple layers of paint. Apply the heat evenly across the surface just until the paint bubbles or rises up. Then scrape off the paint with a plastic paint scraper, using a stiff nylon brush if you're going to ultimately finish the surface natural, or a wire brush if you'll be painting the surface, to remove paint stuck in the recesses of ornate embossing. For stubborn paint caught in the small crevices of ornate tin panel designs, you may need to finish with chemical paint stripper.

Paint strippers are the last resort when it comes to cleaning up a tin panel. You should be careful in selecting a stripper that will not react to the tin plating or the metal underneath (you'll find strippers that are specially formulated for use on metals at your local home center). Brush the stripper across the surface, being generous with the application. Follow the manufacturer's instructions, which usually require that the stripper be allowed to sit on the surface for several hours. Scrape off the paint and repeat if necessary.

SALVAGED PLUMBING PIPE

Plumbing pipe, although plentiful in deconstruction and demolition projects, is more of a challenge to reuse than wood or other metals. Plumbing pipe does not have the original and apparent decorative application that something like a wood floor or tin ceiling would. In that sense, it's not obviously attractive—it requires adaptation. In addition, contractors and deconstruction firms alike tend to recycle pipe rather than try to sell it for reuse. There are many recycling operations dealing in scrap metal, so if you're going to reclaim pipe, you need to get in at the source, or find the rare reseller who stocks pipe.

But that effort can pay off in spades. When and where you do find it, reclaimed plumbing pipe of all sorts is generally very inexpensive. Pipe is also easy to work with, strong, and durable. Most pipe is fairly innocuous in appearance. Copper pipe is the exception. Like other copper fixtures, the finish on copper pipe and tubing can be alluring. But even if you are faced

A wire wheel attachment can be highly effective in removing surface debris from the nooks and crannies of a tin panel's design. But it should only be used on a tin panel that will be repainted, not finished natural.

with using another type of pipe, you can change the look to suit your needs and tastes. Plumbing pipes can be sanded and polished smooth, left natural, or painted any color of the rainbow.

Whatever type you choose, stay away from lead pipe and pipes that were used to carry solid waste. Even though you're not likely to use lead pipe in a project near food, there's too great a chance that the pipe can degrade and flake, creating a potentially toxic situation inside the home. Likewise, sewage pipes carry the risk of contamination.

Use plumbing pipe in a design project and you'll inevitably need to use companion fittings. These include elbows that allow you to turn corners, T fittings that enable supporting connections, reducers, and mounting flanges, among others. These fittings allow you to combine piping into various structures and also aid in mounting a finished project to a wall, floor, or ceiling.

TYPES OF RECLAIMED PLUMBING PIPE

Metal		Profile	Appearance	Best Uses
Copper		By far the most attractive reclaimed pipe. Easy to work with and can be bent without much effort.	Original shiny red finish is eye-catching, but needs to be sealed to keeping from tarnishing; however, patina of age is an alluring matte jade-green. Many people find it preferable.	Joints are soldered because pipe can't be tapped. Use in pot and pan hangers, brackets, towel racks, and more. Not sturdy enough for table legs or other structural support applications.
Iron		Two basic types reused: cast and black iron. Black iron is a bit softer, but both are strong enough even for structural uses.	Cast-iron pipes have a familiar gravel gray appearance and black-iron pipes are black. Both can be sanded and painted, or left unfinished for an industrial look.	Pipes can be used with fittings and elbows to create unique shelving frames, table bases, and other supporting structures.
Galvanized Steel		Not significantly different from iron pipe, for the purposes of upcycling. Very strong and corrosion resistant.	Dull gray-silver natural appearance can be sanded and painted or finished with high-gloss clear finish for a variety of looks.	Strong enough to be used in the same structural applications that cast-iron pipes are.

TECHNIQUES: DESOLDERING COPPER PIPE

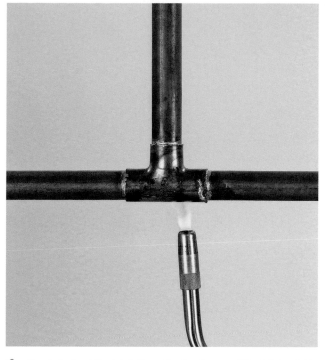

1 Carefully heat the joint. Move the torch around the joint quickly without stopping. Keep the flame an inch or so from the surface of the metal. Attempt to heat the joint evenly.

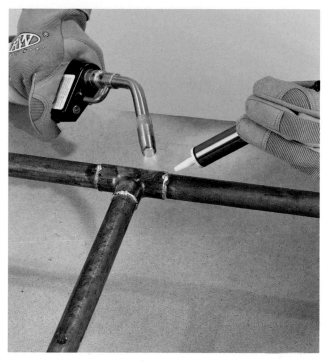

2 As the solder heats, liquefies, and begins to flow out of the joint, suck it up with a vacuum pump (known commonly as a solder sucker). As the flow decreases, you can capture the last of the solder with a soldering wick.

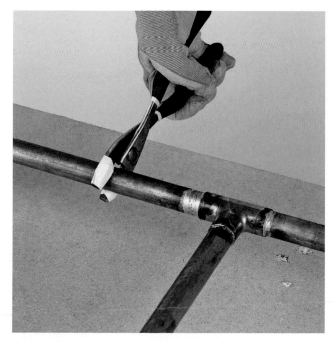

3 Gently and firmly pull the joint or pipes apart using large pliers with rubber grips and jaws that you've padded with masking tape or other protective covering to ensure against marking the metal. If the pieces don't come apart easily, reheat the joint until they do.

✚ SAFETY FIRST

When reclaiming copper pipe directly from an existing structure, you should be absolutely certain that the pipes have been drained of any water. If you heat a copper pipe containing water, the water becomes steam, expands, and can distort or even rupture the pipe. The result can be a bad burn or, at the very least, pipes that are not reusable. Also exercise care when using a torch around a building structure. The heat itself can ignite old and very dry structural members around the plumbing pipes.

CHAPTER 4

RECLAIMING STONE MATERIALS

Stone endures. Even as a building ages and crumbles, the stone components remain unscathed. That's why the many types of natural stone—and their kiln-fired cousins—are prime candidates for salvaging and reuse. The forms that stone and ceramics take are many and varied and, in most cases, incredibly beautiful. They can be reused as decoration, structural elements, or, more commonly, as both. Some, such as carved marble fireplace mantels, are best used as they were originally intended. Others, such as foundation stones or cobblestones, can be reused in many different and even surprising ways. The versatile nature of the material and beguiling range of appearances make stone and ceramics popular reclaimed materials. No matter what your individual home design project may entail, reclaimed stone or ceramic of some sort can play a compelling role.

Stone and its manmade equivalents have been used as building materials for thousands of years. Ancient Romans used decorative stone tile throughout their sumptuous homes. The English built many of their most iconic and lasting structures of locally quarried stone. Early pioneers in the United States built solid, lasting structures with foundations or entire walls constructed of fieldstone. The trend continues to the modern day, with the best buildings including some elements of stone and ceramics.

Such a rich history teaches us that stone is an incredibly useful building material. It can also be incredibly enchanting. As much as builders have been drawn to the inherent strength in stone and ceramics, homeowners, designers, and others have been captivated by the inherent beauty of stone surfaces. There seems to be an endless variety to capture the eye: flecks and veining, colors in every shade, and intriguing surface textures.

There are almost as many ways to use the material as there are variations in appearances. Anywhere you might want or need a durable, engaging surface, some type of stone, brick, or ceramic materials will serve admirably.

USES FOR RECLAIMED STONE & CERAMICS

COUNTERTOPS AND BACKSPLASHES

Stone surfaces and reclaimed tile often come from salvaged countertops, so it makes sense that they would be reused in that role. But even if the materials are rescued from somewhere else in a deconstructed home, they are still prime candidates to serve as countertops. Floor tiles, such as large format (12 x 12 inches) or larger, can easily be repurposed as countertops even if the tiles need to be cut to fit. Terracotta tiles salvaged from a floor or a patio, and ceramic tiles of all sorts, can be repurposed into counters or backsplashes. You may have to do some modifications and special fitting because many older tiles are not standardized sizes. Stone slabs can provide even more dramatic countertop options. However, you may need to cut slabs to fit—a possibly daunting job if the slab is unusually thick. Stone slabs may also need to be sealed before they can be used as countertops in the kitchen or bathroom.

MANTELS

Whether it's a mantel over the fireplace or a decorative mantel along an entry wall, no mantel is more impressive than one made of stone. You can choose from a more precise and formal look of cut stone, such as a granite curbstone, or more visually intriguing raw stone. A stone mantel can be a room's showpiece, but it is also a challenge to mount because of the weight and the fact that the back surface of the mantel has to be completely flat. But with the proper installation and appropriate reinforcement, a stone mantel will last the life of the building.

FIREPLACE SURROUNDS

Stone has long been the material of choice for cladding indoor fireplaces in a beautiful, fireproof, and virtually indestructible material that is also

relatively inexpensive. These structures come in two different forms: carved stone surrounds reclaimed as a single piece, often including hearth, surround, and mantel; and loose stones that are stacked in place—mortared or not—to create a unique and handsome structure. Carved surrounds are rare. They can be impressive and visually dominate a room, but all that design power comes with a hefty price tag. You'll pay for both the stone itself and the craftsmanship. Loose stones, such as ledgestone or stacked flagstone, create a much more rustic and informal look. You'll find a far greater selection of loose stones with which to craft a fireplace surround. But the most plentiful option by far will be reclaimed brick, which is also one of the easiest and most natural materials to form into a surround. And antique brick features so many surprising color variations that you may choose it over other candidates.

HORIZONTAL SURFACES

Stone slabs and stone and ceramic tiles are some of the most sumptuous and wonderful floors you can install. They are best suited to kitchens, but in some cases, a particular stone or kiln-fired material will be perfect for an entryway or even a living room. Reclaimed stone tiles and ceramic tiles often come from newer buildings that have been demolished for one reason or another. Brick and loose stone floors were common in many late nineteenth and early twentieth-century homes, especially in the South. Stone, terracotta tile, or brick floors can be reclaimed in their entirety when you find them, and the aging may have added a wonderfully worn surface. Whenever you're considering a reclaimed stone floor—especially if it's replacing wood or carpet—keep in mind that the surface will be colder underfoot year-round.

PATHWAYS

Just as they make wonderful floors within a house, stones of all kinds make excellent pavers for an unforgettable patio, garden path, a walkway, or even a driveway. Outdoor surfaces are a natural use of any stone or kiln-fired material, because there is an organic connection between the material and outdoor area. Any flat stone can be used in this way: slate, granite sidewalk slabs, and unfinished marble countertop pieces to name just a few. Of course, dimensional pavers such as bricks and cobblestones are also perfect for these types of surfaces. You'll have to choose between an irregular design, such as a flagstone patio, or the more regimented look you'd get in laying a surface such as a herringbone brick pathway.

Stone	Profile	Appearance	Best Uses
Granite	One of the hardest and most-durable stones, and one of the most elegant. Most reclaimed granite is unpolished and unsealed and may need to be polished, sealed, and possibly cut for use inside the home.	Dusty muted shades of light and dark gray. There are rarer light reds, black, and purple granite. Granite has an enchanting speckled coarse to fine grain.	Reclaimed countertops should be used as countertops; curbstones and pavers make wonderful flooring, mantels, and steps; be aware that any reclaimed granite used on the interior of the house should be sealed against stains and water infiltration; polish for elegant high-gloss appearance.
Limestone	A common building stone often used in the past to build barns, walls, fences, and outbuildings because it was easier to cut and stack than milled lumber. Also widely used in commercial buildings in the early twentieth century.	Chalky white look with relatively large open pores; less frequently off-white or light pink; travertine is a sophisticated version of limestone with a mottled surface similar to marble, and in amber-yellow, whites, dusty reds, and black.	Countertops, warm-climate floors or patio surfaces (used indoors, the stone should be sealed); can be stacked to make a fireplace surround, or garden wall.
Sandstone	Formed from sand and silt on riverbanks, most reclaimed sandstone is imported from Europe, where it was used for rural outbuildings. The stone is durable, long-wearing, and ages well.	Intriguing flowing variations and color changes, colors include tan, caramel, and black to pink and terracotta red. Can be finished in any sheen from matte to high gloss.	Countertops, flooring, wall tiles. Used anywhere moisture is present, the stone needs to be sealed.

Stone	Profile	Appearance	Best Uses
Fieldstone & Flagstone	The names come from the fact that these were harvested from fields to build houses, barns, walls, and other structures.	Fieldstones are round and irregular; flagstones are flat and break apart in layers. Both have coarse, earthy surfaces and earth-tone colorings; best used in informal rustic settings.	Fieldstone is more difficult to work with and is usually mortared into place for walls, fireplace surrounds, and fire pits; flagstone is stacked dry to build garden walls, fireplace surrounds, or vertical surface cladding. Both are great patio, sunroom, or pathway surfaces.
Slate	Extremely hard and durable stone. The stone is salvaged from many applications, inside and out, and you'll find both regular, cut pieces, and irregular slabs. Very useful because it is less porous than other stones.	Interesting surface variations including unique veins and unusual grain patterns. Colors range from black and deep purple to dusty blues and grays, to turquoise and light red. Combinations in the same slab are not unusual.	Often used as flooring (cut and uncut), because of the uneven slip-resistant surface; also used for countertops and wall cladding. Large slabs make good tabletops.
Marble	The king of reclaimed stone, marble is a rare salvage treasure that is usually sold—even reclaimed—for a high price.	Reclaimed marble is available, when it's available, in a vast array of colors and patterns, from jet black, heavily veined, to Carrara white marble, to pink, green, and purple.	Most reclaimed marble is already cut, either as countertops, floor tiles, or slabs for other applications. Reclaimed samples can easily be cut for a variety of uses, including flooring, wall surfaces, backsplashes, vanity and kitchen countertops, tabletops, and more.

WORKING WITH STONE TILES & SLABS

Cleaning reclaimed stone must be done carefully to avoid any damage to the stone surface. This is less of a concern with less expensive, coarser stone such as fieldstone, or where you're planning on reusing the stone on the exterior of your house. In any case, start with the least-powerful cleaning procedure, working up to more caustic and potentially damaging materials and techniques. Consult a professional for expensive stone surfaces such as marble.

Reclaimed stone tiles, countertops, and slabs can be some of the best bargains in a salvage yard. The cost of these is often a 30 percent or more reduction of the price of the same materials if purchased new. Of course, there are tradeoffs. Reclaimed tiles, countertops, and other forms may not be standardized thicknesses. Be prepared (as you always should be in working with any reclaimed materials) to deal with irregularities.

You should also be ready to do a bit of prep work to get the materials in shape for the project you have in mind. The greatest cost savings—and the biggest challenges—will come from reclaiming tiles or slab surfaces yourself. Reclaiming antique stone is a laborious process no matter what form it's in. Wall-mounted slabs such as mantels will be removed by deconstructing the wall around the stone, and then carefully prying the stone away from and off of its mounting fixtures. Reclaiming a curbstone or other stone from an outdoor location generally means prying it out of the installation. Existing countertops are removed by first removing any attachment hardware, and then carefully prying the surface free (it helps to have a helper; long surfaces should be pried away from the mounting structure at two or more points simultaneously to ensure against breakage). Reclaiming tiles depends on the method used to lay them. In most cases, you'll need to use a grout saw to remove connecting lines between tiles, and then pry them up carefully with a flat pry bar, taking your time. Resign yourself to a lot of potential breakage.

It's usually well worth the added cost to purchase stone slabs or tiles through a salvage vendor or importer. Not only do you save your labor, the stone will normally have already been prepped to one degree or another, and the selection will be much greater. In addition to traditional reclaimed materials companies, you'll find a lot of stone options at importers, because England and other parts of Europe have a rich tradition of reclaiming stone from centuries-old buildings. Much of that stone makes its way to the United States.

No matter where you got the stone, you will probably need to cut it to the shape you need for your particular project. You will also most likely have to clean the stone. Depending on what type of stone you use, you may even need to seal the surface.

CLEANING STONE SURFACES

Most stone can be effectively cleaned with a basic mix of warm, soapy water. Always use clear, unscented dish soap and lukewarm water. When cleaning reclaimed stone, first try to remove all the dirt with a soft-bristle brush or rough towel. Then test the soap-and-water mix on a discreet area. Once you're sure that the mix will not stain the surface, wash the stone thoroughly and quickly. Use a stiff scrub brush to remove any stubborn dirt or debris. Rinse thoroughly with distilled water. Immediately dry the surface as completely as you can, using a soft, clean towel. Allow the stone surface to completely air dry before working with it.

Stubborn stains usually require a specialized cleaner, especially in porous stones. Different stones require different types of solutions, so the best bet is to contact a local stone worker for tips on cleaning the particular stone, or plan on positioning the surface so that the stain is hidden.

Cleaning goes hand-in-glove with sealing, if you're hoping to preserve the stone surface. A number of stone sealant products are on the market, each intended for a given type of stone. Buy one that applies to the stone you're reclaiming, and use it according to the manufacturer's instructions to ensure the longevity of the stone's appearance.

A thorough cleaning is the first step in reclaiming stone for home-design projects.

TECHNIQUES: CUTTING GRANITE OR MARBLE SLABS

1 Mark the cut line on the stone surface by centering painter's or colored masking tape over the intended cut. Use a metal straightedge and utility knife to score the cut line over the tape. Remove the tape along the waste side.

2 Using an angle grinder equipped with a stone-cutting blade, cut along the line defined by the tape. The color should guide you even as the grinder kicks up dust and debris. Carefully follow the cut line to complete the cut.

TECHNIQUES: CUTTING STONE WITH A WET SAW

1 Straight cuts across the surface are easily made with a wet saw. Mark the cut on the waste side with an indelible marker. Guide the stone through the saw—equipped with a stone-cutting blade—to make a clean, straight cut.

2 Make edge cutouts by scoring the three inside borders of the cutout with a chisel or carbide tip marker. Cut each edge of the cutout, and make several parallel cuts in from the outside edge to the inside edge, resetting the fence each time. Snap the pieces out of the cutout with pliers.

(Continued)

3 Clean up the cutout by positioning the tile so that the blade is sticking up inside the space of the cutout. Carefully move the tile sideways so that the blade barely skims the inside edge, cleaning up the jagged portions.

4 Make cutouts inside the field of a tile by marking the cutout and positioning the stone underneath the saw blade with the arm up. Make a plunge cut along one side of the cutout, until you almost reach the top and bottom marks. Rotate the tile and continue making plunge cuts to remove the center section.

+ SAFETY FIRST

Using a wet saw can be very dangerous because the saw blade is moving fast and the presence of water makes the tile very slippery to handle. Work slowly and position the fence and tile so that the part of the tile you're moving is farthest from the blade. Use safety eyewear as well, to both protect your eyes and ensure you can see as clearly as possible. When cutting with a grinder or polishing a cut-stone face, use an approved dust mask and eye protection.

5 Polish stone tile cut edges by equipping a grinder or power drill with a polish disc or wheel. Start polishing with an 80- to 100-grit wheel, and work sequentially down to 400 grit, until you have a perfectly smooth surface.

TECHNIQUES: CUTTING SLATE

1 Thoroughly clean the slate and determine which side will be the top. Use a straightedge and marker to mark the back of the tile with the cut line. Set the tile face down on a sturdy, flat, and level surface.

2 Mark along the cut line using a sharp, pointed cold chisel, nailset, or awl. Use a hammer to tap several points about a ¼" apart along the cut line. Use firm but light blows. The chisel only needs to mark the surface and create fracture points along the line.

3 Sandwich the marked slate tile, bottom up, between two scrap 2 x 6s. The edges of the wood should align over and under the cut line, leaving a tongue of slate hanging out on the waste side. Tap at the cut line with a maul or stone mason's hammer.

WORKING WITH LOOSE STONE

Stone has many faces, and although the finer varieties such as travertine, slate, and marble tend to get the most attention, they are hardly the only types commonly reclaimed. In fact, whether you're deconstructing an old barn or strolling the aisles at a salvage company, you are far more likely to come upon less prestigious loose stones. These include fieldstones and flagstones unearthed from local meadows and excavations, and pavers either reclaimed when sidewalks or old streets are redone, or left over from backyard pathway projects. Although they aren't as prized for their appearances as their quarried cousins might be, loose stones are less expensive, more plentiful, and easier to work with.

They can also be easy to reclaim. Flagstone was often dry stacked, which means no mortar holds the stones in place. Cobblestone and other pavers are usually laid in a bed of soil and firmed down with sand—so they can be salvaged with little more than a shovel and some hard work.

The original uses reflect the rugged nature of loose stone. They are the workhorses among natural building materials, the stuff of which foundations are often made. You'll find loose stones holding up barns or comprising the entire walls of much older structures. They were also widely used as dry-laid property walls separating one area from another. Occasionally they were pressed into service as flooring, but there are too many more impressive modern options for that to be the case today. However, loose stones were often used to create fireplace structures, which remains one of the best uses for reclaimed loose stone.

Working with these materials is easier than adapting finer stone surfaces to your purposes. A material such as fieldstone is forgiving of imprecision, and granite pavers can be used without the fine cuts you would need for a marble countertop. The most

Flagstones are plentiful, inexpensive, easy to work with, and ruggedly handsome.

laborious part of working with loose stones is usually lifting, cleaning, and prepping them for reuse.

Although they tend to look rustic, loose stones aren't just for log cabins and lodges set deep in the woods. A stacked flagstone fireplace surround is impressive in just about any style of home. It can also be used for low divider walls and, in recent years, has seen exposure as fronting for accent walls. Pavers can be reused to make outdoor paths, but they can also make unique floors for a bathroom or impressive steps in a mudroom, walls, or even used to make fireplace surrounds. Indeed, the stones are adaptable far beyond their pioneer roots; it's just a matter of looking beyond the rugged nature of the material.

RECLAIMING KILN-FIRED MATERIALS

It is the natural materials, the hand-hewn, vividly grained timber and darkly captivating black slate, that draw the most attention when a building is torn apart. But human-made pieces, in the form of kiln-fired tiles and bricks, can be every bit as alluring and useful. In fact, some of the most dramatic decorative elements you might rescue from a building are those that were born in a fire.

Kiln-fired materials, such as building brick and terracotta tiles, are likely to be some of the most structurally intact pieces of a building (as long as

the process of deconstruction itself has not resulted in their purposeful destruction). Bricks have been used for centuries in this country because of their structural integrity, fire resistance, and natural beauty. In Europe, terracotta has been included in buildings dating back a millennium or more. And still they endure.

Bricks are the most commonly reclaimed kiln-fired materials. Bricks became popular as a structural building material in the early 1900s, due primarily to their innate fire resistance. Because they were so easy to build with and are structurally sound and impervious to burning, they were used widely in valuable properties such as factories, textile mills, and large barns as well as public and commercial inner-city buildings meant to stand the test of time. They were generally used to form the walls, but builders didn't stop there. Certain types of bricks were used to pave streets, walkways, and private driveways in many parts of the country. Although the vast majority of these surfaces have either been dug up or paved over, the bricks from many of them live on in salvage company warehouses.

Many of the antique bricks you will reclaim, or that are available on the market, were manufactured by hand. The exact processes differed region to region and manufacturer to manufacturer. This makes for a lot of variation in reclaimed brick size and appearance. The bricks were also often marked with a manufacturing stamp, such as the name of the maker or, in the case of many antique bricks manufactured in Chicago, the words "union made." Colors range from the familiar bright red to black. Certain colors are indicative of a type of manufacturing or use. For instance, antique fire bricks that were used to line the fireboxes of residential fireplaces are often yellow or cream colored, with fascinating variations as a result of long exposure to extreme temperatures. Textures are similarly varied. Brick can be found with rough course surfaces or nearly smooth faces. These bricks have often lasted 100 years or more, and could quite possibly last another 100.

Terracotta tiles are a fairly recent home decor addition in the United States, but they have been a common fixture in the floors, countertops, and patios of Europe for more than 1,000 years. That's why most of the reclaimed terracotta tile available now is imported from Europe. It's every bit as enchanting as you might expect of a material that adorned Roman villas. Just like antique brick, the firing method and clay composition used to make the tiles influences how they look. Bits of straw and other debris are sometimes apparent, but not in a detrimental fashion.

The physical appearance of these tiles has been tempered by their long use, abundant sun, and the occasional wine spill. The finish, coated as it is with the residue of countless dinner fires and the bottom of sandals, has an intriguing visual depth. The colors range from creamy off-white to yellow, dusty red and deep purple to almost black. The tones are exceedingly rich.

Antique terracotta tiles are even less regimented in format than old brick is (although many suppliers cut them down to uniform sizes). The tiles come in a variety of distinctive shapes, but the most prevalent are square, rectangular, diamond, and octagon. The actual size of the tiles varies not only place to place, but also in tiles from the same location. They will differ slightly in outside dimensions, but can differ even more in thickness, because both the manufacturing and laying of the tile were done completely by hand. Some were stamped or etched with insignia, such as initials or family crests, while others were marked with standard designs of the time, such as

These antique square European terracotta tiles are rich with a history that is reflected in the well-worn surfaces and variegated colors. The tiles are well suited for duty as flooring, backsplashes, and even tub and tile surrounds.

Antique ceramic tiles were often painted to form intricate and enchanting large-field designs.

Terracotta roofing tiles are often reclaimed and can be used in many decorative ways.

fleur-de-lis. The styles, size, and coloring also differ greatly depending from country to country, and even within a given country. For instance, antique terracotta tile from Italy's Tuscan region is visibly different from antique tiles reclaimed from the vicinity of Rome. Terracotta roofing tiles are also sometimes reclaimed, but their use is limited given the unusual arcing shape.

Regardless of where they come from or what appearance they boast, terracotta tiles must be carefully installed and sealed. The surface is prone to problems such as efflorescence if not properly treated. However, most suppliers of terracotta tile also supply formulas for treating the surface, as well as detailed instructions for proper installation.

Reclaimed ceramic tile is somewhat easier to work with, although significantly more fragile. Truly antique varieties can be spectacular examples of handcrafted artistry and are rarities. Tiles from early America as well as those from Europe—dating from the 1600s to 1900s—are one-of-a-kind pieces that carry the price tag and allure of other true antiques. Given the expense (a single tile can run hundreds of dollars), these are usually only used as accents in a decorative tile scheme that won't see use as a work surface. However, salvagers regularly reclaim ceramic tiles of more recent vintage. But even though the tiles may not be hand painted, they do often feature a patina of age, such as whites that have mellowed into a creamy yellow, and many include machine-stamped designs that will have degraded in intriguing ways.

TYPES OF RECLAIMED BRICK

Before brickmaking became a largely mechanical and automated process, the quality and character of any given batch of bricks varied in the extreme. In addition, different bricks were—and are—used for different applications. Bricks that would comprise the substructure of a building did not need to be weatherproof or particularly attractive, while the face brick that formed the facade of a building like a bank had to be both. Early manufacturing processes were variable. The look and integrity of any given batch of bricks depended on the quality of materials (which could change given irregular supply), the abilities of

the workers, the speed at which the bricks needed to be produced, and even where the bricks were positioned in the kiln. But all bricks—reclaimed or new—are broken down into general classifications according to the intended use.

COMMON, OR BUILDING BRICK

As the name might suggest, common bricks—also known as building bricks—were meant as workhorse units. In general, they are not considered the most beautiful bricks and appearance was secondary to structural performance. Although intended for functional uses, common bricks were often used as facing brick when a lack of facing brick might have held up construction of a building. This was generally done on cruder buildings where appearance was not a major concern. Used in exterior facing applications, many common bricks would fail; spalding was a regular problem. However, defective bricks are not usually reclaimed, so the common bricks rescued during renovations or deconstruction are likely to be sound. More modern types of common bricks include frogged, which are manufactured with indentations, and perforated, which can have many small or a few large holes through the brick. These are lesser forms and are usually not reclaimed.

FACE BRICK

Antique face brick can be a valuable reclaimed material because color, texture, and size are all more uniform than with other types of antique brick. The face brick was usually the best of any given lot and was manufactured to stand up to harsh weather and still maintain its attractive appearance. The look is usually indicative of a given region, and these can be some of the most beautiful bricks.

PAVING BRICK

These bricks were specifically formulated to withstand the abuse of traffic and the elements. If you're considering putting in a pathway, patio, or driveway, you should always select bricks designated as pavers. Common bricks will generally not hold up to freeze-thaw cycles in areas of the country where the ground freezes, and you should never use a frogged or holed brick for paving purposes. Pavers were also made to resist moisture infiltration. They were used as the road-paving material of choice from the introduction of the automobile in the first decade of the twentieth century through the development of asphalt in the 1930s. In fact, quality paver bricks are the equal to granite cobblestones.

There are many different types of antique brick pavers available, from standard rectangles to interlocking keystone bricks. Reclaimed and salvaged pavers often have a lovely worn smooth side. The colors are deeper than modern bricks, and antique bricks do not fade over time or under the glare of sun exposure. Like most bricks, the look, size, and shape of antique pavers are usually unique to the region in which they were made and laid. Many companies offer pavers by geographic designation, such as Chicago pavers.

CLINKER BRICK

Clinkers are some of the most unusual, interesting, and downright fun bricks to use. Formed in the hottest areas of the kiln, their location resulted in odd colors and often distorted shapes. Clinkers were at first considered waste bricks and discarded. They were named for the distinctive, almost-metallic sound they made when thrown onto the waste pile. But in the 1920s, craftspeople realized that these bricks were special and could be used in special ways. Dark purple and black bricks were incorporated into the face of walls on commercial buildings to spell out the initials of the company that owned the building. Misshapen clinkers were included in exterior walls to create unorthodox and incredibly interesting designs. Clinkers are, by the nature of their production, harder, heavier, and denser than other bricks, making them very resistant to moisture and wear. They can be used in walls, benches, and pathways, but they also add something special to fireplace surrounds. And the exacting tolerances of modern manufacturing methods ensures that the existing clinkers are all that will ever be made.

FIRE BRICK

Although in a more general sense, firebricks are any that are used to line ovens, kilns, and fireplaces, reclaimed firebricks are the ones used to line the fireboxes of residential fireplaces. These were specially designed to withstand extremely high temperatures. The manufacturing process created many captivating colors, including cream, gray, red, yellow, tan, and black. Exposure to open flame over decades or a century or more added scintillating depth to these colors, altering the hues and creating patinas that can't be found in any other brick type.

RECLAIMING & CLEANING BRICK

You can reclaim brick from exterior walls or interior locations such as a fireplace surround with just a little hard work. Use a pneumatic hammer with a blade bit on the mortar joint between bricks. Older brick was mortared with limestone mortar, which is softer than modern cement mortar. Work from the top down to remove bricks, and expect that 10 percent of the bricks may be destroyed in the process. Don't concern yourself with removing all of the mortar while you're excavating bricks from a wall—that is part of the following cleanup process.

Removing bricks from a mortared location raises a lot of dust and can make a lot of noise. Always wear appropriate eye protection, ear protection, and a dust mask. To dull the vibration of the pneumatic hammer, it's wise to wear a quality pair of thick work gloves. Work boots are also a good idea, because the bricks tend to fall before you can catch them.

As with other reclaimed materials, cleaning up salvaged brick is a sequential process in which you use the least severe methods first, moving on to the more serious processes only when you have to.

Physical removal of mortar is the place to start. Lime-based mortar is often degraded enough to make removal less of a chore than it might otherwise be. A long soak in a strong vinegar solution can also loosen mortar on bricks in many cases. You can use pure vinegar but be aware that the same dangers (albeit to a lesser degree) exist with a strong vinegar solution as with any other acidic solution.

If all else fails and you are considering the brick for a purely decorative purpose, you might consider using a wet saw or grinder to cut the most attractive face off the brick. This type of brick "veneer" has been used throughout history to create the illusion of a brick wall over other surfaces. You can use it to the same effect in installations such as fireplace surrounds.

Chisel large sections of mortar off the brick, using a small cold chisel with a fine blade and a hammer. Secure the brick in a vise with padded jaws, and work slowly and carefully to avoid damaging the brick.

Use a power drill equipped with a wire wheel attachment, or a wire wheel on a bench grinder to remove the remaining mortar debris.

SALVAGE WISDOM

Stubborn mortar residue can also be removed by soaking the brick in an acidic mortar-removal formula available from hardware stores, or a 10-percent muriatic acid solution (1 part acid to 10 parts water). However, muriatic acid is an extremely dangerous substance (a form of hydrochloric acid) and you must use all safety precautions recommended by the manufacturer. Start by removing as much of the mortar as possible with a chisel and hammer. Make sure you wear all necessary safety gear, including rubber gloves that go as far up the arm as possible, acid-resistant safety eyewear, and a respirator. Work outside or in a room with very strong ventilation. Set up an acid bath in a plastic or glass tub and carefully add the acid into the water—never the other way around. Then soak the brick with water before placing it in the muriatic solution. The acid usually bubbles as it goes to work on the mortar. Allow the acid to work on the mortar for about 10 minutes, then use a stiff scrub brush to brush the mortar off the brick, brushing away from your face and body. Neutralize or otherwise remove any chemical formulation you've used on the brick once you've removed the mortar. Baking soda can neutralize many acidic formulations—you just add the baking soda to the mix. Once neutralized, rinse the solution off the brick thoroughly with copious amounts of water and scrub it with a scrub brush and a mild dish soap and water solution. Allow bricks to dry completely before using them.

CHAPTER 5

RECLAIMING GLASS

In no other building material is the difference between old and new more marked than it is with glass. A timeworn mirror will have lost some of its "silvering" and gained some charm, intermingling the reflected image with scars from the lost backing. Antique windows proudly bear the liquid imperfections caused by earlier crude manufacturing methods; the swirls, bubbles, and striations are endlessly engaging. A rescued beveled cabinet door lite bears elegant witness to a handcraft that is dying and disappearing. In all its many forms, reclaimed glass is something special, beautiful, and inevitably unique.

Glass is like no other reclaimed material. Although it doesn't play the key structural role wood and stone can, it does have the capacity to radically alter the building it goes into just as it did the one it came from. As a replacement for a more energy-efficient solid surface, it allows light and air to enter a building, and provides access to a view. Other materials may take many shapes and fulfill many roles. Salvaged glass comes in a limited number of forms. The most common, of course, are windows. Fortunately, windows themselves take many different shapes, sizes, configurations, and styles. Reclaiming an older window can entail no small degree of work, but the reward is often an architectural feature that serves as a standout in the modern home.

Extracting an older window from an existing structure is a delicate operation, but not technically difficult. The idea is to remove the frame holding the pane or panes carefully enough that you do not break any of the glass. The difficulty of doing this successfully depends on the type of window and where it's located. You'll also encounter glass as part of a door—either an interior or exterior door, or as cabinet door inserts. Removing this glass for reuse is a great deal easier than reclaiming an entire window, frame and all. Mirrors are glass too, but are most often used in the same decorative fashion they served in their original location. But no matter where you're reclaiming the glass from, the first step is always to judge whether the effort is worth it.

REUSABLE OR NOT?

Judging whether architectural glass is prime for reuse instead of recycling involves a number of different factors. Start with the goal you have in mind for the glass you're considering. For instance, if you're looking to put a reclaimed window into an existing opening, the reclaimed window needs to be the correct size for your opening, or you'll need to modify the opening. If you're considering a window for an exterior exposure, you'll likely be doing some work on the window to ensure its insulation values are as high as possible. That's why reclaimed windows are ideally suited to an interior partition wall.

However, if the structure of a window is compromised by rot, water damage, or insect infestation, pass it up unless there is something special about the glass itself. In which case, you can extricate the glass from the sash and frame, and discard the damaged wood. In addition, not all reclaimed windows are antique. Many double-insulated and other modern styles come on the market when newer homes are remodeled. Before you buy one of these, be sure that the seals on the window are intact, and that neither the glass nor the frame is cracked. Also ensure that the window precisely matches your opening, because there is no way to resize a double-insulated window (although you can reframe the window to fit an overlarge opening).

Glass inserts in cabinet doors can also be a real find, and don't necessarily need to front an existing kitchen cabinet. Instead, if the door is distinctive enough, it may be worth making a standalone cabinet to show off the glass, or retrofitting existing shelving to accept the reclaimed door. On the other hand, if the door rails and stiles are not in good repair and

the glass is plain, undistinguished window glass, it's probably not worth your time. You can pull decorative glass, or glass art, out of a cabinet door and reuse it to front an existing cabinet door. This is quite a simple process: one of the great things about reclaimed glass is how easy it is to resize—it's just a matter of a quick and easy cut.

MIRROR IMAGE

Windows have their allure, but antique mirrors can seem magical. Many are sold with the original frames intact, although here again, you should be certain that in buying a framed mirror you aren't importing wood-eating insects into your home. Inspect any frame closely. The back of the mirror is also worth a close inspection. If the backing seems to be chafing off, it could be toxic and a hazard to have in the home.

Keep in mind that although a mirror may seem part of a whole with its frame, the mirror can be used in many ways apart from the frame. Always consider a mirror in light of how it would look frameless. You can easily reframe a mirror, or hang it without a frame. And, of course, you can use a flat mirror as a cabinet door insert, a small tabletop, or in many other applications. In the case of a highly decorative frame, the frame itself may serve a wonderful design purpose apart from the mirror. A hand-wrought tin or gold-leafed frame might make a wonderful border for a piece of art or a large photo.

A salvaged mirror with eroded silvering may offer an appearance even more intriguing than the image it reflects.

SALVAGE WISDOM: RESILVERING MIRRORS

Sometimes, a reclaimed mirror may have lost so much of its silver backing that it is more a dilapidated mess than an enchanting antique. When that's the case, consider resilvering the mirror.

Many manufacturers offer silver stripping and resilvering products, and some even offer complete turnkey kits. Using these products is not a difficult process, but it has to be done carefully and correctly for an attractive appearance. Be aware that the process is very messy. Make sure you have plenty of space to work.

1. Remove the mirror from its frame, taking care not to scratch the mirror. Assess the glass condition. Flaws or damage to the face of the mirror will be accentuated with new silver.

2. Strip the backing paint. You can do this with a plastic fid, or use any of the chemical stripping products available on the market. Follow all safety precautions recommended by the manufacturer. In any case, always wear safety goggles, a respirator, and rubber gloves.

3. Strip the silver. Use a mirror removal product, following the manufacturer's instructions. This usually involves applying the product and then removing the silver (and possibly copper) backing with gauze or cotton balls.

4. Pour tin over the glass, let it sit for about 40 seconds, and pour off any excess. Rinse with distilled water. Mix the silver according to the manufacturer's instructions, and then pour the silver evenly over the surface of the glass. The surface will turn brown, and then silver. After five minutes, drain any excess silver off the surface of the glass.

5. Let the surface dry completely. Remove any traces of silvering on the face of the mirror by carefully scraping them off with a plastic fid. Reframe the mirror.

When cleaning the face of any antique mirror, spray a lint-free cloth with window cleaner and clean the window with the cloth. Never spray cleaner directly on the face of the glass.

RECLAIMED WINDOW TYPES

There are many different types of windows. Some have historical precedent, and others are modern innovations. Most are defined by the way they are opened and closed (or not) and each can serve a different decorative reuse. The window you choose will depend on exactly what project you have in mind and what type is available.

FIXED

Many—if not the majority—of reclaimed windows are this style, which consists of a basic frame, often with muntins separating different panes into what are called *divided lites*. Operable windows required technology that was simply not available to the builders of many early twentieth-century homes and business buildings. To owners of industrial buildings throughout the nineteenth and early twentieth centuries, windows that opened were an unnecessary expense. But as plain as this style may sound, there is still a good deal of diversity within the category of fixed windows. For instance, antique fixed windows were regularly manufactured with frames of non-wood materials. Copper and other metal frames make for a very interesting look that can accent other decorative elements in your interior design. Homeowners who use these windows often leave the frames in their original condition, covered with the patina of age. Even framed in wood, an older fixed window can feature a lovely distressed look. The simplicity of a design with no moving parts makes these windows ideal for reuse as interior windows, tabletops, and even picture frames.

SINGLE- OR DOUBLE-HUNG

These common styles are still used today. Featuring top and bottom sashes in which one or both slide up and down (single or double respectively), the windows can easily be taken apart so that each sash can be used separately. A fashionable trend in recent years has been to use one sash with divided lites as a wall-mounted frame for photos. A sash can also be used to create a shadow box for collectibles or other items you want to display. Older single- and double-hung windows were not standardized. The size you find on a deconstruction site or at a reseller's shop may be unique and not well suited to replace an existing window. However, many newer models in standard sizes are becoming available as late twentieth-century homes are being taken down and reclaimed. Because of the hidden mechanism of counterweights, these

windows are difficult to reclaim yourself. It's wiser to buy them from a salvage source.

CASEMENT

Casement windows open from one side just as a door does and are usually a single pane of glass. This type of window is less popular than others for a number of reasons, and it's a hard window to reuse for anything but a direct replacement of an existing failed unit. A steel-framed casement window can be separated from the jamb and opening unit to be used as a coffee tabletop, but other reuses in home design are limited.

AWNING AND HOPPER

Generally smaller in size than other windows, awning and hopper windows are intriguing styles that can, in certain instances, change the look of an interior. The awning window opens from bottom out like an awning, while the hopper is just the opposite—hinged on the bottom to open like a hopper drawer. These styles can be used to great effect as over-door transom windows in a home. They can be used purely as decoration, or as conduits for extra light and airflow throughout the home. Awning windows are sometimes installed as interior wall windows, kept closed most of the time, but open when air circulation is needed. Either style is sometimes used over a curio or shadow-box table or cabinet, where the opening mechanism allows easy access.

JALOUSIE

The jalousie window was originally intended as a tropical style that could open to allow air circulation and at the same time prevent rain from entering through the window. The window is actually a set of horizontal glass louvers that are cranked open and shut. The style is associated with mid-twentieth-century architecture, and retrofitting an existing window opening with a jalousie window can add flavor to the look of a home, both inside and out. Frosted-glass versions are wonderful replacements for bathroom windows because they can be opened for ventilation while maintaining privacy. Jalousie windows are most often used as replacements because of their architectural interest, but they are sometimes mounted on walls for an interesting sculptural effect. With a little imagination, the intriguing form of a jalousie can be used in home design in a number of different ways.

EYEBROW

Eyebrow windows are arched designs that are usually fixed but occasionally operable. The interesting shape makes a wonderful interior transom window, one that is appropriate for many different home styles. It's also a great choice for a window in a partition wall. Antique versions often feature elaborate frames, such as starburst or spiderweb designs, dividing the window up into tiny segments of glass. The frames are sometimes iron or other metal, adding to the allure of the window. When the frame is ornate, distressed, or otherwise distinctive, eyebrows can be fascinating decorative features all by themselves. Simply set atop cabinets or even bookshelves, the window draws attention. The shape lends itself to interior installation, such as at the top of a wall bordering an interior room with no exterior exposure.

PORTAL

Portals are some of the most unusual windows you can reclaim. Completely round, these windows are often salvaged from ships, and they are consequently framed in brass, copper, or other metal. But even framed in bent wood they are enchanting. The round shape brings an uncommon element to any interior, and portal windows with detailed frames and attached fixtures such as hinges or latches are often displayed on a wall all by themselves. Non-operable portal windows look incredible when installed in a solid-core door or, in a set, along an interior wall. The round shape also lends itself to a second life as a table. Find one with antique or textured glass in the opening, and the table will likely serve as a conversation-starting centerpiece. Because they have traditionally been used more on ships than in houses, you can find the largest selection from firms specializing in marine salvage.

SALVAGED ART GLASS

Decorative glass treatments have been part of this country's history since the very first buildings were constructed. Stained glass has been a staple of churches not only here but throughout the world. But that's just the tip of the iceberg. The Victorian period saw an explosion of both plain leaded glass windows and stained glass in homes and other buildings. Beveled glass designs have been around for almost as long. More recently, sandblasted glass has played an increasingly prominent role in buildings both residential and corporate. Consequently, you may find decorative art glass in a building under deconstruction, but there is a huge selection among salvage firms and antique dealers. Given the range of styles that are available, you should be able to find a piece of art glass that's just right for your home with just a little searching.

LEADED

The term *leaded* is somewhat misleading, because most art glass—stained and beveled included—is technically "leaded." But in most cases, the phrase refers to the fact that individual pieces of glass within the window's borders are held in place by lead channels (they have

a U or H profile) called *came*. Leaded windows are made of plain glass cut into decorative shapes and assembled into an overall design. The lead channels are soldered together to hold the design in place, and then anchored inside the actual window frame. In general, this means that leaded windows can't be resized. But depending on the design and how much it excites your inner designer, creating a space to accommodate a leaded-glass window can be well worth the effort. The designs themselves range from simple squares and rectangles to much more ornate shapes and figures. Leaded-glass windows come in all shapes and sizes because they were, until fairly recently, custom made. They are not installed in operable windows, but you can replace a fixed window that is larger than a leaded-glass unit by building out a frame in the opening to accommodate the smaller leaded-glass window. Some designs are so detailed and eye-catching that people mount the window on a wall by itself, although they can also be used as glass fronts for cabinets or intriguing tabletops.

STAINED GLASS

Few reclaimed architectural artifacts are as steeped in history as stained-glass windows. The art of stained glass has been around for centuries, although the windows that you're likely to reclaim will probably be a few decades old at best. Regardless of age, stained-glass designs range from kitsch to sublime. Even if you're not a fan of the stereotypical intensely figural designs that most people think of when they think of stained-glass windows, take heart. You can find purely geometric treatments and very simple designs that are nothing more than a combination of squares and rectangles in one or two colors. The colored glass itself is fascinating. Crafted in almost every color imaginable, stained glass does not fade or wear away—the pigment is an integral part of the glass. However, vintage stained-glass windows—especially any used in exterior applications—can literally fall to pieces. If you find one you like, check for bowing or gaps between the lead channels and the glass pieces, and inspect for any cracks or checking in the glass. A modestly worn stained-glass window is easy to repair, but more extensive damage calls for professional help and may rule the window out. In addition to actual stained glass, you'll also find samples of painted glass windows. Artists have long painted on stained glass as they would on a canvas to create more lifelike scenes than could be crafted with the glass colors alone. The effect can be marvelous, and light passing through a painted stained-glass window will be muted in dramatic dusky jewel tones.

Used in the right location (one that receives significant direct natural light), a painted window can be stunning. But both stained and painted windows are best used in moderation; a single small window can create a showpiece. They can be installed in existing window openings just as a leaded window would be. A stained-glass window can also make a stunning work of wall-mounted art if placed in a frame with backlighting. Smaller panels are vibrant additions to kitchen cabinets or transoms. The windows can also make amazing tables sandwiched between two pieces of tempered glass and edge banded with wood or metal to form a flat surface.

SANDBLASTED

Sandblasted, or frosted, glass is a more contemporary version of art glass. Artists use thick blast-resistant stencils to etch designs into windows, glass walls, or onto mirrors. As with other forms of art glass, sandblasted designs can be cheesy, sophisticated, and everything in between. The decorated surface permits light to shine through, while serving as a privacy screen. Solid frosted windows and other glass surfaces can be reused as short partition walls, tabletops, counters, and sidelites for interior or exterior doors. Sandblasted windows really come to life when the window can be viewed at different angles and is lit by both artificial and natural light sources.

BEVELED

The art of beveling an edge onto a thin piece of glass takes skill and well-calibrated machinery. But before technology and automation became prevalent, beveling was strictly a craft that required talent and a breadth of knowledge. Most antique beveled windows were made by hand, and the craftsmanship is blatantly obvious. Craftsmen lay an edge angled to make the most out of light passing through the glass. The right bevel creates a brilliant sparkle and a prismatic effect that breaks sunlight apart into a rainbow of colors.

Floral designs from the early decades of the twentieth century complement fussier architecture, such as Victorian homes. More geometric designs provide a pleasing linear simplicity that is appropriate for modern, contemporary, traditional, and craftsman-style interiors. Beveled inserts are at home in exterior windows where the play of sunlight creates color bursts throughout the day, and in cabinet doors, where the natural sparkle improves the look of whatever is kept inside the cabinet.

Like other glass art, many beveled designs are held together with lead channels that are subject to the stresses of age. This is something to consider before putting a beveled panel in a structurally stressful location such as the door lite for a front door.

VINTAGE SPECIALTY GLASSES

Study antique glass panes closely and you'll probably find intriguing imperfections. Bubbles, thick areas, and streaks are all part of what was once a very imprecise production process. They are also flaws that are beautiful and desirable. But as glassmakers became more proficient and exacting in how they made glass, they realized they could control the process to make uniform the flaws that were once accidental. The result were specialty glasses, a variety of textures that were either purposely imprinted on the molten glass surface to create stunning, clear glass patterns, or manipulated into the form of the glass to create body and life in the forever frozen material. Originally used primarily to create privacy without reducing light transmission, these glasses have become decorative pieces in their own right.

TEXTURED

The process is fairly simple and mechanical, which is why textured glasses are sometimes referred to as *machined* glass. Right after the hot sheets of glass are formed, a steel form or stencil imprints the surface with a texture. The texture can be random or, more often, it is regular and standardized. Textured glasses are still made today, but many of the patterns available through salvagers are unique. Panes of textured glass are most at home where privacy is needed, such as a bathroom door or window. But they can bring exciting touches to the doors on cabinets of any type. It's a way of having glass-fronted cabinet doors without worrying if what's in the cabinet is worth showing. Textured glass is fairly easy to cut, so panes can usually be resized to fit any opening.

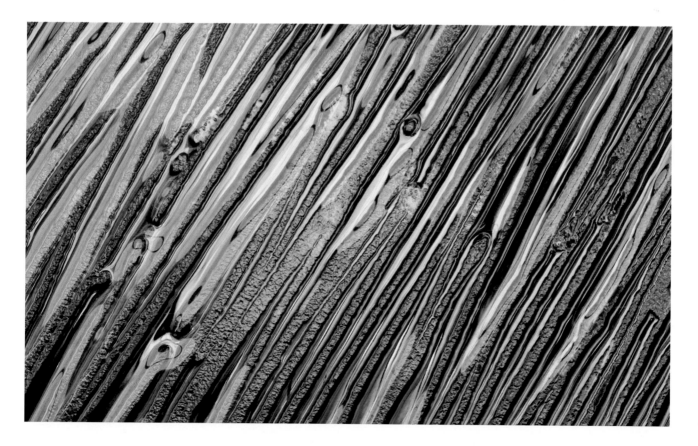

SPECIALTY GLASSES

The same companies that produced stained glass used a similar process to form clear specialty glasses in a range of styles. These are often named for the visual effect, such as *water glass* that looks like flowing water, and *seedy*, which looks as if clear seeds have been embedded in the glass. The effects are widely varied and the patterns are often unique. Depending on what form the glass was given, it may entirely obscure a view, partially obscure, or simply add wrinkles to what you see. Like machine-textured glasses, these types can be cut and resized for use as tabletops, replacement windowpanes, door and sidelites, and more. They can even be painted white on one side and used as backsplash tiles.

INDUSTRIAL GLASS

Chicken wire glass is a specialty glass that was regularly used in industrial settings. The glass is formed with wire inside it to prevent shattering, and the surface was often textured as well. The glass was specially made in the ages before tempered glass to add security and ensure against breakage. The design also prevented injury; if broken, large pieces of sharp glass would be held in place by the wire. Because it was used primarily in industrial settings, chicken wire glass is often framed in sturdy, industrial metal framing. The framing itself can be visually arresting, and the combination makes for a focal point replacement window. Chicken wire glass is almost exclusively available in fixed-window form and can't easily be cut down to size or modified. But large panels are used as room dividers, shower doors, or windows. Some are sturdy enough to be used as dining room tables or desks. Corrugated glass is similar to chicken wire glass, in that it was used in factories as a strong and safe transparent surface—usually for awnings, roofs, or skylights. Corrugated glass panels are generally large, with the rippled form you would expect from the name. The panels can be professionally cut down to form fronts for breakfast bars, half-height divider walls, and other sculptural uses.

1 Remove all the old putty from around the panes of glass and remove any cracked or broken panes entirely. You can use linseed oil applied and left for an hour or so to soften old putty, or use a heat gun. Move the gun slowly and continually along the putty line.

2 Using a putty knife, pry out the old putty as soon as it softens. Scrape the glass channel clean, removing all the old putty and dirt, but take care not to gouge the window's wood. Use needle-nose pliers to pull out any glazing points left in the channel after you've removed the putty.

3 Sand lightly as necessary to clean out any stubborn putty. Once the glass opening is completely cleaned, prime the surface for the wood putty by brushing on a coat of linseed oil or primer. If the wood isn't sealed, the dry surface will draw moisture from the putty, compromising the final glass seal.

4 Roll a thin bead of putty to about the diameter of a chopstick, and line the channel. Smooth it in place with your thumb.

(Continued)

5 Clean the glass—especially the edges so they will seal with the putty—and lay the pane in the putty bed. Use gloves to avoid cuts and smears on the glass, and press the pane lightly but firmly down. Wiggle the pane slightly from side to side as you press down, until it is completely seated.

6 Drive glazier's points into the wood frame to hold the pane in place. Use the tip of the putty knife to slide the point into position against the pane and into the wood frame. Use at least two points on each side of the pane.

7 Roll a rope of putty and press it along the seam between wood and glass. Push the putty down into place and create a neat beveled edge by dragging along a putty knife held at a 45° angle. Remove any excess putty.

8 Let the putty dry for several days until it is no longer cool to the touch. Prime and paint the sash, and over the putty bead. Paint onto the glass to ensure a good seal and scrape off the excess paint with a razor.

TECHNIQUES: CUTTING RECLAIMED GLASS

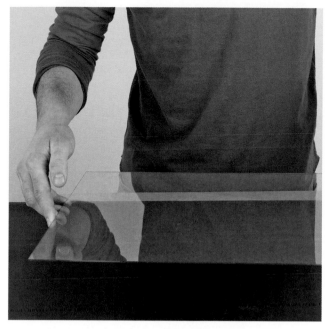

1 Clean the glass thoroughly. Place the glass on a clean, flat, and level surface. The surface should be rubber or another material that will prevent the glass from sliding as you apply pressure. Textured glasses normally have a smooth side, opposite the textured side. Cut the smooth side.

2 Mark the cut line along the glass. Use a metal straight edge, one that is thick enough to avoid catching the cutter's wheel, and align it along the cut line.

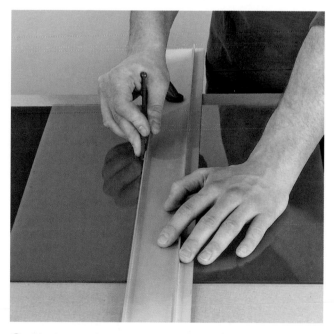

3 Dip the cutter's wheel in light oil or fill the oil reservoir. Tuck the handle between the index and middle fingers, with the point of your thumb bracing the underside of the handle. Hold the cutter at a slight angle and press down with firm pressure. Draw the cutter along the straight edge toward you in one smooth motion.

4 Hold the pane of glass along the edge of a table or scrap piece of wood so that the cut line is aligned with the edge of the underlying surface. Place a hand on either side of the glass as you would hold a newspaper; snap both ends down. If the glass doesn't easily break, tap along the cutting line with the breaking ball on the end of the cutter's handle.

CHAPTER 6

SALVAGING FIXTURES & ARCHITECTURAL ACCENTS

Old houses, barns, other outbuildings, and workshops can yield fascinating period fixtures and detailing along with the wood, metal, stone, and glass of the structure. These smaller, self-contained accents can be big finds. An antique sconce, intricate wrought-iron bracket, or an older farmhouse sink can all serve as centerpieces in a room redesign. Accents and fixtures are generally reclaimed and resurrected to fulfill their original functions, but they can also be used in new and innovative ways.

Fixtures and accents salvaged from buildings being demolished, updated, or simply falling down of their own accord usually feature styling specific to a given period. But that doesn't mean you have to be a historian or know exactly what era a doorknob or light fixture comes from to incorporate it into your house. It comes down to what your tastes are and how the feature will work with the other elements in your home design. Luckily many different salvageable items can complement just about any decor or home style.

Cut-glass doorknobs, simple art deco wall sconces, or a cast-iron tub that is just the right shape and size to suit your bathroom remodel will all work regardless of what the rest of your home looks like. One of the wonderful things about many architectural accents is that they can stand on their own, just as a painting or photo would.

Reclaiming some of these will be a matter of need. After all, you don't just decide to swap out your bathroom sink and vanity because you happened upon a hand-painted ceramic unit from the 1940s. But in other instances, you can shop with an eye to possibilities. If you have a house full of darkly stained doors, you might be open to the box full of brass doorknobs and locksets you find at a local antiques shop or salvage warehouse.

TYPES OF SALVAGED FIXTURES & ACCENTS

PLUMBING FIXTURES

The vintage tubs and kitchen sinks you'll find on the reclaimed marketplace are not only uniquely styled, but they are also usually made of almost indestructible enameled cast iron. Every historical period has its own style and the same is true of sinks, tubs, and toilets from those periods. Sinks are available in all the shapes you'll find today, but most are surface mounted or standalone. Original farmhouse sinks with the front apron style that is popular in today's market are a wonderful find. Older pedestal sinks can feature incredibly stylistic flourishes, such as fluted column bases, multi-tiered edge forms, and units with formed backsplashes. Both kitchen and bathroom sinks were made of porcelain and porcelain-plated cast iron. A smaller number of kitchen sinks were made of stainless steel, usually under a brand name created by the manufacturer. You'll find vintage wall-mounted and pedestal bathroom sinks and older vanity sinks. Kitchen and utility sinks were manufactured as standalone units as well, but with single or multiple bowls. Many featured very cool and useful built-in features, such as

soap dishes and drainboards with grooves that direct fluids back into the basin.

Of course, there are also newer sinks on the market, reclaimed from houses built in the 1960s or from remodeling projects. These can include stone sinks such as marble, and versions made out of engineered or synthetic materials. As a general rule, if you're going to the trouble of shopping for a reclaimed sink, the biggest reward will come in the form of an iron, metal, or high-end stone sink. Synthetic versions are less likely to hold up over time or feature distinctive styling.

Reclaimed bathtubs are available in much the same materials as period sinks are, but they come in two basic styles: apron or built-in, and freestanding (pedestal and clawfoot). Salvaged built-in tubs often feature elegant styling and shapes. Standard apron tubs slide right into a U-shaped cavity, while alcove tubs have a wraparound apron that allows them to sit in the L formed by two intersecting walls. Alcove tubs can only be positioned one way, fitting into either a right-hand or left-hand corner. Clawfoot and pedestal tubs are freestanding and can be positioned anywhere the drain can connect to the waste line. The tub can be serviced by faucets or fixtures routed up through the floor or out of a wall. Clawfoot models are some of the most impressive antique tubs and will work with many different styles of home decor. Although pedestal tubs are somewhat more rare, they are impressive in their own right. They require a lot of floor space, but can make a powerful design statement in the right bathroom, not to mention being a luxurious place to take a bath. Many older tubs are exceptionally long or deep (or both) and provide a sumptuous bathing experience. However, these variations in size are why you need to measure carefully to ensure that you have the proper space to accommodate a reclaimed tub.

Toilets are usually reclaimed only if you're trying to re-create a true period look. Older toilets were generally six-gallon units that don't conform to modern codes or best environmental practices. Reclaimed toilets can, however, be incredible bargains if you can use them in a bathroom renovation.

Plumbing hardware includes faucets, showerheads, decorative handles, and the exposed faucet bodies used with freestanding tubs. These are stylish, often crafted in brass, steel, and copper. Their appearances, whether left aged or cleaned up to sparkle like new, separate them from the more common contemporary chrome fixtures. Faucets, spigots, and showerheads are usually rebuilt with new valves, screens, and washers. Some parts, such as the pretty ceramic handles so often re-created by modern manufacturers, can be reused for purely decorative purposes.

CHAPTER 7

SECONDHAND STUFF PROJECTS

Reclaiming, reviving, cleaning, and mending salvaged building materials is only the first step in building with secondhand stuff. The fun part, and the more rewarding stage, is using that material to craft new and useful creations. Some reuses are simply that—reusing the material as it was originally intended, such as flooring that is cleaned up and reused as flooring. But more often, salvaged material can become something else. A pallet becomes a stunning modern coffee table. A door becomes a handsome headboard for your bed. The projects in this chapter give you lots of options for the materials you reclaim, but they should also be considered departure points for your imagination. With a little bit of ingenuity, you can find your own unique uses for materials from the past.

Optional: For a distressed finish, scatter sharp rocks over the island counter top and put a scrap piece of plywood over the rocks. Walk on the plywood to scar the wood. Finish the top with an antique finish (right) or, if you'll be preparing food on top, use a food-safe finish, such as boiled linseed oil.

17 Place the top in position and screw it in place by driving 3" wood screws up through the nailers.

PALLET PROJECTS

There are two ways to reclaim and reuse pallets: deconstruct them and build with the loose lumber, or use them largely unmodified. The projects here are examples of what can be done with the pallet as-is. Not only are they both useful and attractive, they require less work than taking a pallet completely apart.

The first project, the pallet coffee table, is the simplest version of pallet furniture. For this project, the pallet should be a single surface pallet—one that has a layer of boards only one side (or you can simply remove all the nonessential, structure-retaining boards from one side). Although this project requires a fair bit of sanding and painting, it is exceedingly simple, and even the novice DIY will find it an easy Saturday afternoon project. The result will be a high-end modern look that is subtle enough to fit into a contemporary or traditional interior design.

The pallet planter is a much-beloved project for gardeners of all types, because it allows any gardener to make good use of vertical space. The planter requires few materials and is inexpensive to craft, and can easily be mounted to a wood, plastic, or chain-link fence, a wall of the house, or even just leaned up against a garden wall. The planter looks best when planted with a combination of bushy and trailing plants, and it works wonderfully as a space-conserving kitchen herb garden.

HOW TO MAKE A PALLET COFFEE TABLE

A pallet coffee table is an elegant piece for any decor.

(Continued)

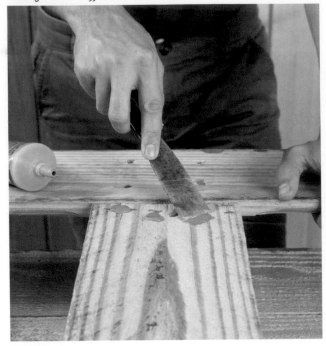

1 Clean the pallet and remove any nails sticking out. Use sandable wood putty to fill any depressions or hammer marks, and cover nail or screw heads.

2 Sand the pallet all over with an orbital or palm sander. Prime the pallet, and then paint it with a top coat of gloss latex white paint, being careful to avoid drips.

3 With the pallet laid upside down on two sawhorses, screw 4 x 4 x 2" caster blocks at each corner, securing them with 4" screws.

4 Position a 6" swivel-plate caster at each corner. (At least one should be a locking caster.) Drive 2" wood screws through the caster's flange holes to secure them in place. Measure the top of the table to the edges on all sides and order a glass top, or multiple glass pieces cut to fit, with edges beveled. Lay them in place and dot the top with self-adhesive clear rubber bumpers where the glass lays over a board. Flip the glass and it will be secure where you lay it.

HOW TO CREATE A PALLET PLANTER

1 Clean the pallet and remove or pound down any protruding nails. Leave the two end boards and one middle board. Remove boards with a claw hammer or pry bar, or use a long-handled pallet buster. If you're mounting the planter on a fence or wall, mount it now. Be sure the fasteners are strong enough to support the added weight of moist soil and plants. Paint the pallet after it is mounted, if you are painting it.

2 Measure and cut a sheet of thick landscape fabric about 1' longer than the width of the pallet and three times as wide as the cavities between the boards. Fold the edges in, and the front and back over, and staple them to create a pocket as deep as the width of one individual pallet board.

3 Slide the fabric pocket behind one front board of the pallet so that it sits securely in the cavity. Make any necessary adjustments. Staple the landscape fabric pocket in place. Repeat with the remaining board cavities. Fill the pockets with potting soil and plant your preferred plants. Water thoroughly. For even more visual interest and to remember which plants were put where, you can staple a plant identifier tag, seed packet, or screw on a label holder in front of the plants. You can even draw on the board with chalk or colored pencil.

Indoors or out, a pallet planter is a vertical organizer for all kinds of greenery.

HOW TO INSTALL A TIN BACKSPLASH

The space between countertop and cabinet bottom is often a wasted home design opportunity. Because you don't actually use the surface, it's all too easy to overlook the design potential of a backsplash. But this particular area is ideal for cladding in reclaimed tin. It's a modest-enough surface that the tin won't overwhelm the kitchen, but the exposure is visible enough that you'll get a lot of bang for the buck from this particular design accent.

The techniques used here are incredibly simple, and the materials and tools needed are all quite basic as well. You should be able to put up a tin backsplash in the space of an afternoon without breaking a sweat. What's more, this same process can be used to put up tin wainscoting, clad a pantry door, or even cover an accent wall in reclaimed tin panels. We've used construction adhesive to hold up the panels

here, which should be fine for your kitchen as well. However, if you want to take extra precautions, you can nail the panels to studs, nailing through the nailing rails on the panels. One caveat though—in dealing with cut tin, you'll be working with very sharp edges. A moment of inattention can result in a fairly serious cut. Always use the thickest work gloves you can find, and be careful when manipulating panels into position.

You'll also want to seal both the surface and seams of the backsplash. The kitchen is home to many spills and splashes, and any moisture that makes it underneath the backsplash panels can lead to rust and, potentially, mold. That's why you'll seal all the edges with clear caulk, and seal the surface in a clear finish or high-gloss paint.

These reclaimed metal ceiling panels make a beautiful and comment-generating kitchen sink backsplash.

1 Measure the backsplash area, including the location of any wall switches or outlets. Look at the embossed design on the tin panels and determine where you want them positioned, and what overall pattern you want to create along the length of the backsplash.

2 Lay the tin panels out on a clean, flat work surface to determine their final positioning. Mark the layout lines on the face of the tin panels for cutting, using a black marker and a metal straightedge. Mark the position of all wall switches or outlet cutouts. Cut the first panel to fit.

3 Make the cutouts for switches or outlets by drilling a large hole at one corner of the area to be cut out. Use tin snips (inset) to widen the hole, and then cut along the cutout line and remove the cutout.

4 Coat the back of the first panel with construction adhesive in a serpentine pattern. The pattern should distribute adhesive evenly over the surface. But keep adhesive away from the panel's edges to avoid any squeeze-out during installation.

(Continued)

5 Position the panel in place on the wall, and press it down. Make sure the entire surface is firmly in contact with the backsplash surface.

6 Coat the back of the next panel with construction adhesive and place it into position. The panels should overlap at the edge nailing rails. Be careful not to spread adhesive onto adjacent panels.

7 Secure the panels in place by taping them together with painter's tape or duct tape. You should leave the tape in place until you're absolutely sure the construction adhesive has fully set and the panels are stable.

8 Caulk around the edges of the panels using tinted (available from aftermarket tin ceiling manufacturers) or clear caulk, in the seam between the tin and the countertops, and between the tops of the panels and the bottoms of the cabinets. Lay a bead of caulk along the edges in corners as necessary to ensure a watertight seal for the backsplash. Paint if desired.

HOW TO MAKE PIPE COAT HOOKS

1 Clean up the plumbing pieces you'll use for each hook. (Each requires matching flange, 2"-long threaded nipple, elbow, and square head plug.) Scrub the pipes and fixtures with steel wool if painting and use metal paint. All the pieces must be the same diameter.

2 Assemble the hook by screwing the nipple into the flange, the elbow onto the other end of the nipple, and the plug into the end of the elbow. Screw the hook directly into a stud, wood wall, or into anchors in drywall, using flathead screws sized to match the hole, material, and finish of the flange.

Simple, attractive, and useful, pipe coat hooks are a stylish reclaimed project.

HOW TO MAKE A PLUMBING PIPE TOWEL RACK

A towel rack such as this is easy and quick to assemble, is customizable to your desired length, and has a cool industrial look.

1 Clean up the plumbing pieces you'll use for the rack: at least a 2 x 24" threaded pipe, two matching elbows, two 2" long nipples, and two flanges. Use a wire wheel to remove any old finish, corrosion, or grime on the surface of the pieces.

2 Screw the elbows onto each end of the pipe and tighten them as much as possible. Screw the other end of the nipples into the elbows and tighten completely, while still aligning them.

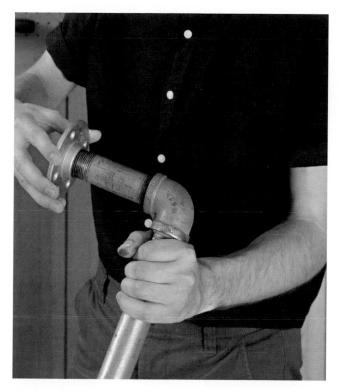

3 Screw the flanges onto the free ends of the nipples on each side.

4 Prime and spray-paint the towel rack in your desired color. Black and white are the most common colors, but many different colors are available. Keep in mind you can easily repaint the rack after it has been assembled.

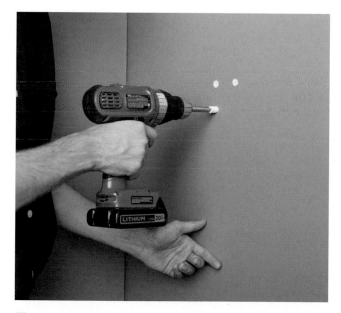

5 You'll need to use anchors to mount the towel rack. Hold the rack in position and mark through the holes in the flanges. Remove the rack and drill the holes for the anchors. Screw or tap the anchors in place.

6 Check that all the connections are tight before mounting the rod. Screw the flange bases into the anchors, using screws sized to match the hole, material, and finish of the flange. *Note: You can use longer nipples so that the bar projects farther out from the wall to use as a hanging coat rack. Or use a longer pipe to make a curtain rod.*

HOW TO BUILD A GRANITE-SLAB KITCHEN ISLAND

Countertops are some of the best uses for reclaimed stone slabs. Not only are the sizes and shapes of many slabs ideal for this application, some stone surfaces were countertops in their previous incarnation. The combination of a luxurious appearance, outstanding durability, and resistance to wear and tear (including physical abuse and water) make all kinds of stone ideal for food preparation surfaces in the kitchen.

Although we've chosen to describe adding a reclaimed stone top surface to a rolling cabinet that will serve as a mobile kitchen island, the same techniques described here can be used to adapt reclaimed stone as countertops over built-in cabinetry. Attaching the top works the same in both cases, with masonry screws holding the stone surface to the bottom structure. In both cases you may need to cut the stone you've selected to the correct size, or to make room for a sink or other feature, such as a cooktop.

How you finish the stone surface is a matter both of what type of stone it is and what type of appearance you're after. Choose a slab of slate and it's possible you can seal it without polishing for a matte black or gray surface. Opt for granite, and you'll probably want to seal and finish it in a high-gloss coating that is not only protective, but makes the intrinsically interesting grain really pop. Keep in mind, though, that many types of stone need to be maintained if used as kitchen surfaces. Both granite and marble may need to be resealed every 6 to 12 months, depending on how you've sealed them to begin with, and how they were finished.

Ultimately, though, no matter which type of reclaimed stone or finish you choose, the surface is bound to be an opulent addition to the kitchen. There is a richness to true stone—especially antique stone—that can't really be replicated in synthetic materials. Go to the effort of reclaiming a stone surface and integrating it into your kitchen's design, and you've added an unrivaled accent for a fraction of what it would cost new.

After

1 Remove the cabinet drawers, shelves, and all hardware for cleaning. If the hardware is in poor condition, or you find it unstylish, replace it. Check the cabinet's structure to determine if it needs reinforcement to support the stone top.

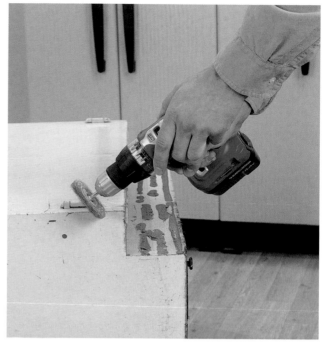

2 Prepare the cabinet for painting or finishing. Clean, sand, or just brush off rust or loose paint as necessary. Wipe down the cabinet with mineral spirits and a rag before you begin painting it.

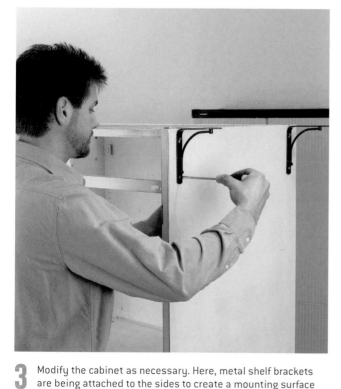

3 Modify the cabinet as necessary. Here, metal shelf brackets are being attached to the sides to create a mounting surface for the countertop. The brackets add decorative appeal. Use decorative hex-head bolts to attach the brackets. If your cabinet doesn't have a back panel, add one.

4 Prime the cabinet carcass, doors, and drawer fronts with metal primer. You'll get the best results if you use a sprayer, such as this HVLP sprayer, in a well-ventilated workroom. Make sure the primer goes on as a thin, even coat.

(Continued)

5 Paint the cabinet. For best results, apply several thin coats and use a paint formulated for metal (if your cabinet is metal). Sand lightly between coats to remove any imperfections, according to the paint manufacturer's instructions.

6 Install casters on the bottom of the cabinet carcass so that it can be moved around the kitchen and repositioned as needed. Alternatively, you can make the cabinet into a kitchen island by anchoring it to the floor.

7 Complete the cabinet. Reattach the hardware or install new. Add decorative drawer pulls for an eye-catching accent. Reinstall the cabinet doors, making sure that they close correctly and hang level and plumb.

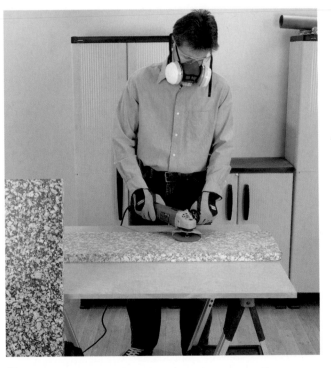

8 Clean up and rehabilitate the salvaged countertop. Here, two 12"-wide strips of 1½" thick granite are seamed together to create the countertop. Before installation, the top surface and edges of each strip are polished with an angle grinder equipped with a diamond wheel. Wear a respirator, eye and ear protection, and work in a well-ventilated area.

9 Position the countertop strip on the cabinet with the overhangs roughly equal at the sides. Apply a bead of clear polyurethane sealant to the mating edge of the first strip.

10 Place the second strip on the cabinet and press it firmly against the first. Use bar or pipe clamps to hold the strips together.

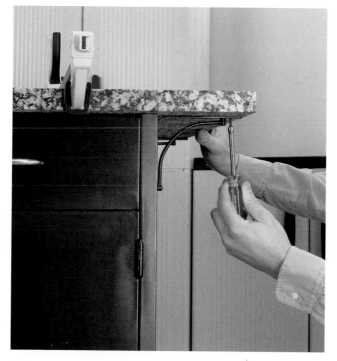

11 Anchor the countertop to the mounting brackets. Use a masonry bit to drill holes in the underside of the countertop to accept masonry screws. Apply a few drops of epoxy into each screw hole before fastening the brackets to the countertops.

SALVAGE WISDOM: STONE ODDS AND ENDS

Small projects such as this cabinet top are ideal uses of reclaimed stone. Although you may sometimes find large sections of reclaimed granite or marble that will provide all the surface area you need for an entire kitchen's worth of countertops, more often than not, you'll be faced with smaller sections and mismatched pieces. You can use these for islands, or incorporate them as an accent section as part of a longer run of countertops in a complementary material such as Corian. You can also cut and polish pieces to make large cutting boards or incredibly beautiful sideboard hot plate trivets.

HOW TO BUILD AN OLD-WINDOW GREENHOUSE

Old, discarded windows are just about the most obvious choice for reuse in upcycled greenhouses. As homeowners across the country upgrade to insulated windows, older windows are relegated to the junk heap. These include wood-framed units that have seen better days, aluminum-framed storm windows that are no longer needed, and even vinyl-clad insulated windows that have come to the end of their lifespan. You can find them in dumpsters, piled up on remodeling job sites, for sale cheaply through salvage companies, and online through various sources—often for free.

The trick to reusing these windows for a greenhouse is that you have to take what you get. Sizes vary, sometimes radically. The design shown here is typical, featuring a small footprint and a simple shed roof. The recycled windows used are 28 inches wide by 62 inches high; the larger windows are used in the walls, and the smaller are used in the roof. In both cases, these instructions assume wood-framed units. Wood framing is always best when it comes to adapting old glazing to a new hobby greenhouse, but you have to make sure that the wood is in good shape. The lumber used is all of common sizes and should be readily available through salvage companies or from job sites. The design can be scaled up or adapted to different sizes of windows fairly easily, as the framing details have intentionally been left somewhat crude to allow adaptation without a lot of fuss. A kneewall base ensures that no matter what windows you reuse, they'll be well supported. We've also used fiberglass panels, which are widely available and simple to install, for the roof.

When creating a design such as this, using mostly recycled materials, it's a great idea to lay the windows and framing studs out on big, flat work surface—for example, a garage floor or expanse of lawn. If you're not necessarily good with the nitty-gritty of detailed measurements and nominal lumber sizes, this can be a way to verify that everything will fit and to make quick adjustments without having to tear things apart. Whatever the case, it's always

Build a greenhouse of old windows and you not only save money, but you also create a distinctive structure that puts a dumpster-worth of waste materials to great use.

wisest to try to find multiple windows in the same size; cobbling together windows of odd sizes throughout the structure can lead to a nightmare and create an unstable and unsafe greenhouse.

That said, any upcycling project involves certain compromises. This greenhouse will not be as airtight as most polycarbonate panel or plastic-film kit greenhouses would be (and certainly not as much of a high-quality custom structure), so it is at best going to serve you as a three-season structure—it just wouldn't be cost effective to heat it over a cold winter. You can make the structure a little more efficient by recaulking the window panes in wood-framed windows and covering smaller openings or gaps between the windows in the roof and the wall top plates with plastic.

TOOLS AND MATERIALS

Cordless power drill and bits	Construction adhesive	Jigsaw
Claw hammer	Caulk gun	Tamper
Paintbrush	6d and 8d nails	Paint or stain (optional)
Sealant or paint	Finish nails	Eye and ear protection
Stepladder	Circular saw	Work gloves
Clear silicone caulk	12" spikes	

1 Inspect all of the windows for significant defects you might have missed during the reclaiming process. Look for rust, excessive warping, and hidden insect damage or rot (if you're using wood-framed units). Clean up the windows and sand or use a wire wheel to remove loose or flaking finishes or rough areas. Paint the frames with primer or coat with preservative. Fasten the 4 × 4 frame together by joining the timbers at the corner with 12" spikes, countersunk. Measure and mark the trenches for 4 × 4 foundation timbers. Dig them 4" deep and fill with crushed gravel. Tamp the gravel down all around.

2 Check the frame for level on all sides, side by side, and diagonally. Add or remove gravel under the foundation to level any one side.

3 Working on a clean, flat surface such as the lawn here, lay out and build one side wall. Nail the sole and top plates to the studs after ensuring that each stud is at a 45° angle to the plate with a speed square.

4 Raise the wall, check for plumb, and brace it into position. Nail the sole plate to the foundation timber with 8d nails.

5 Once the two side walls are constructed and nailed in position, build and raise the rear (non-door) wall into place and nail it to the foundation 4 × 4.

6 Screw the walls together at the corners with 3" wood screws.

(Continued)

7 Screw 2 × 4 × 28 ¼" spreaders in the stud cavities of the side and back walls. Toe-screw the spreaders on the inside edges and face-screw through the corners (inset). The tops of spreaders should be 32" above the 4 × 4 base, faces flush with the outside edges of the wall studs.

8 Screw 2 × 4 × 28 ¼" spreaders in the stud cavities of the side and back walls. Toe-screw the spreaders on the inside edges and face-screw through the corners (inset). The tops of spreaders should be 32" above the 4 × 4 base, faces flush with the outside edges of the wall studs.

9 Use a jigsaw to cut notches in ⁵⁄₄ cedar decking to use as window sills on top of the spreaders.

10 Dry-fit the cedar sills and then miter the corners if the fit is correct. Coat the tops of the spreaders with construction adhesive and nail the cedar sills in place with exterior finish nails.

11 Miter-cut retainer strips from ¾" quarter-round stock for the window openings. Drill pilot holes for nailing, then coat the bottom faces of the strips with construction adhesive and nail them in place with exterior finish nails. You may have to adjust for the windows you reclaim. The windows here varied in height by up to ½".

12 Lay a thin bead of clear silicone caulk into the retainer strip channel for the first window.

13 With a helper, carefully and evenly set the window into position. Have the helper hold the window in place for the next step.

14 Pin the window in place with small brads to allow the silicone to set. Repeat the installation process with the rest of the windows for the three walls.

(Continued)

15 Measure and cut 2 × 6 fascia boards for the top of each wall. Use 3" 6d galvanized nails to tack the board to the top plates of the wall all the way around. The fascia should cover any gaps at the tops of the windows.

16 Measure and cut 1 × 3 cedar strips to run between the fascia at the top and the sill. Drill pilot holes and nail the strips in place on the studs with 3" 6d nails.

17 Measure and cut ½" exterior grade plywood for the half-wall cladding at the bottom of each wall. Nail it in place with 2" 6d nails.

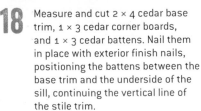 Measure and cut 2 × 4 cedar base trim, 1 × 3 cedar corner boards, and 1 × 3 cedar battens. Nail them in place with exterior finish nails, positioning the battens between the base trim and the underside of the sill, continuing the vertical line of the stile trim.

19 Frame the front wall (12" higher than the side walls) in place by measuring and marking stud and door-jamb positions and toe-nailing the members in place. Nail the top cap to the top of the studs and install the door header. Install the door so that it opens outward.

20 Use 2½" deck screws to toe-screw the front cripple studs for the triangular top walls on both sides.

21 Measure and cut the other cripple studs in each top wall and toe-screw them in place with 2½" deck screws. Check that each is plumb and square to the top plate. Miter each end of the top plates and screw them to the cripple studs on each wall.

22 Measure and cut the 2 × 4 rafters and toe-screw them running front wall to back wall. There should be a 6" overhang on both ends

23 Measure and cut the 2 × 2 purlins for the gaps between the rafters. Toe-screw them in place, spaced 11⅛" on center.

24 Cut and screw the fiberglass panel nailing strips to the top of each purlin.

25 Drill pilot holes for the fiberglass panels and screw them in place to the purlins using 1" screws and rubber washers. Paint or stain the exterior of the greenhouse and, if using in a cooler season, cover all the clerestory windows with 6 to 8 mil plastic sheeting stapled in place.

(Continued)

HOW TO BUILD A RECLAIMED WINDOW CABINET

There are many uses for reclaimed windows in the home. Although the most obvious is simply to install an antique window into an existing opening, you might want to turn to more creative options. Few uses are as charming and practical as the window cabinet described here. It's not just the allure of a distressed or obviously older window. Although that can certainly be a draw, the real attraction of this particular design is that it is so adaptable. The construction of the cabinet body is easy enough that anyone with basic woodworking skills and some fundamental tools can put the whole thing together in about a day. You can also use just about any window you find (round windows won't work) and alter the dimensions to suit the window. A small transom window, for instance, could be used for a bathroom toiletries cabinet. Use two windows to make a dual-compartment cabinet. A long, thin window can be positioned sideways to make a low-slung media center cabinet. And the options don't stop at your choice of window.

We've used reclaimed lumber for this carcass, something you might want to consider if you want to carry through the whole reclaimed ideal—and if you want to create a unique look in the wood carcass that supports the window. Of course, whether the wood is reclaimed or you use some pieces you have lying around the garage, you can choose a finish that suits your tastes and decor. Paint the window frame and box a bright color to fit in with your retro living room, or finish it all-natural to serve as a bar cabinet in your contemporary dining room. Between paint, stains, and clear finishes, your options are mind-boggling.

And don't forget the accents. For the cabinet shown here, we chose exposed hinges and handles that carry through the antique style. You can select copper,

Before

After

bronze, chrome, iron, or other hardware for just the right finishing touches on your own cabinet.

The number of shelves is up to you as well. Depending on what you intend to store in the cabinet, you may want more shelves, or you may decide to position shelves differently. We've placed one right in the middle of the cabinet and cut a slot for the shelf that mirrors the dadoes we cut for the top and bottom of the cabinet. But you could just as easily use any of a number of shelf support systems on the inside of a cabinet carcass. Many of these allow for adjustability, which again, might be your preference. And like the number of shelves, you can decide to build a freestanding cabinet as we have here, or a wall-mounted unit. Although you can clad the back in a simple ¼" plywood sheet cut to fit the outline, we've left the back open. If you leave the back open, you can hang the cabinet by screwing it to top and side cleats that are attached to the wall or, if you add a back, you can screw the back directly to the wall. There are many other ways to hang the cabinet—a trip to the home center will yield many different possibilities. But whether you hang the cabinet or let it stand all by itself, it's sure to be an exceptional focal point in your home's design.

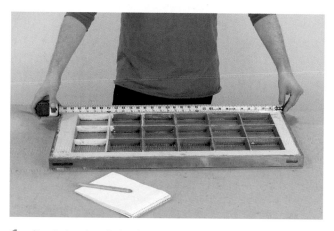

1 Check that the window is square. Remove the hardware, clean, and re-putty the window as necessary. Measure the outer dimensions of the window and note them.

2 Mark and cut the planks that will serve as the sides approximately 4" longer than the height of the window. This is to account for the dadoes that will hold the top and bottom, and to create top and bottom lips. If you prefer, you can cut the sides to the height of the window, and butt-join the top and bottom pieces to the sides.

3 Cut the top, bottom, and shelf for the unit. In this case, they each must be 1" longer than the width of the window. This will allow for ½" of each cross piece to rest in the dadoes on each side.

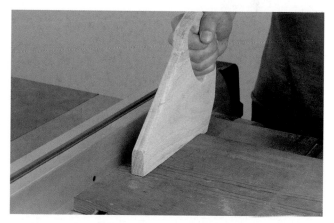

4 Carefully measure and mark each side plank so that the height of the window is centered along the length of the plank. The top and bottom dadoes will be cut just outside these marks. Use a table saw or router to cut ½" dadoes on the least-attractive faces of the planks.

(Continued)

5 Assemble the cabinet carcass by sliding the top, bottom, and shelf into the side-plank dadoes. Drive finish nails through the outside faces of the sides, into the crosspieces. Use a nailset to sink the nails and putty over them.

Alternative: If the cabinet will need to support heavy weight or will be hung in a high-traffic area, use screws instead of finish nails. Pre-drill countersunk holes and attach the pieces with flathead wood screws. Plug over the screw heads with wood plugs of the same type of wood.

6 Sand and paint or stain the cabinet body and window in whatever finish you've chosen. If you're leaving reclaimed wood in its existing state and finish, seal the puttied nail heads and any bare wood surfaces with clear polyurethane.

7 Attach the hinges to the window, and position the window in place. Mark the hinge holes on the cabinet side and door hand latch on the other. Pre-drill the holes and install the window on the front of the cabinet. Hang the cabinet if desired.

HOW TO BUILD AN ACCENT DOORKNOB COAT RACK

This eye-catching coat rack can establish the style for an entryway or mudroom, and the design here can be easily modified to suit your own tastes, or the salvaged goods available to you.

1 Cut the molding or casing down to the measurements that suit your reuse plans. Clean and sand the crown molding section as necessary. Prep, prime, and paint it, or apply whatever finish you prefer.

2 Set the mounting piece on its back and play with the positioning and arrangement of the doorknobs. Measure and mark the location of each doorknob.

(Continued)

3 Check that the dummy spindles you've purchased fit tightly into the doorknobs. Hold each spindle in place and check the orientation of the doorknob if it has a design that reads differently with the doorknob in different directions. Drill the holes for the dummy spindles' mounting screws. Tighten the spindles down, making sure their orientation is correct.

4 Slip the rose for each doorknob down over the spindle with the spindle centered in place in the rose opening. Screw the roses to the molding.

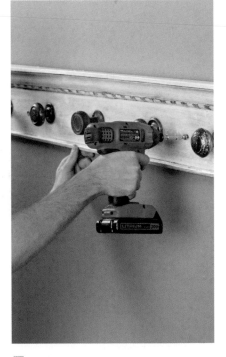

5 Lay a bead of silicone caulk into the rose and smear caulk onto the spindle. Slide the knobs in position onto the spindles and let dry.

Optional: Use stain over the white paint to create an aged, distressed look.

5 Drill countersunk holes for the coat rack mounting screws (unless you're using some other hanging system). Screw the coat rack to the wall and putty or cap the screws, sand smooth, and paint or finish the coat rack.

RESOURCES

BUILDING MATERIALS REUSE ASSOCIATION

A nonprofit educational and research organization with the goal of encouraging efficient building deconstruction and the use of reclaimed building materials.

(800) 990-2672

www.bmra.org

BUILDING GREEN

Organization that offers information, newsletters, and other resources on building green, including reuse of reclaimed building materials.

(802) 257-7300

www.buildinggreen.com

THE DECONSTRUCTION INSTITUTE

An organization providing information and tools exploring and explaining deconstruction as an alternative to demolition and adding to the waste stream.

(941) 358-7730

www.deconstructioninstitute.com

THE ENERGY & ENVIRONMENTAL BUILDING ALLIANCE (EEBA)

A resource providing information and education about sustainable and environmentally responsible building practices and an advocate promoting sustainable construction.

(952) 881-1098

www.eeba.org

FOREST STEWARDSHIP COUNCIL (FSC)

Independent nonprofit dedicated to responsible management of the world's forests. Provides certification of wood that has been responsibly managed and/or represents environmentally responsible wood products for consumption.

www.fsc.org/about-fsc.html

HABITAT FOR HUMANITY RESTORES

Information about the nonprofit's retail outlets selling reclaimed building materials.

www.habitat.org/restores/default.aspx

REUSE DEVELOPMENT ORGANIZATION

A nonprofit dedicated to promoting reuse as an environmentally responsible, socially beneficial, and economical way to manage solid waste.

(410) 558-3625 ext. 15

www.loadingdock.org/redo

US GREEN BUILDING COUNCIL

Nonprofit organization promoting green building practices and socially responsible waste management—including salvaging, reuse, and reclamation—through information and programs. The council administers LEED (Leadership in Energy and Environmental Design), the standardized, internationally accepted green building rating system.

www.usgbc.org

INDEX

abatement procedures, 22
accent doorknob coat rack, 155–156
acetone, 43
adhesive wood putty, 45
alcove tubs, 103
aluminum, 64
American chestnut, 29
antique ash, 29
antique oak, 31
antique elm, 30
apron sinks, 102–103
architectural accents/ornaments
    about, 101–102
    reclaiming, 25
    types of, 102–105
    art glass, 93
    asbestos, 22, 24
    ash, antique, 29
    awning windows, 91

backsplashes, 72, 136–138
bathtubs, 103, 107
bay laurel, 29
beams, 34, 53–54
bench grinders, 109
beveled glass, 94
bi-fold doors, 53
board-and-batten doors, 51–52
brass, 109
bricks
    about, 80–81
    for floors, 73
    reclaiming and cleaning, 84–85
    types of, 82–84
    building bricks, 82–83

cabinets
    doors for, 52–53
    hardware for, 104
    reclaimed window, 152–154
    reviving vintage, 56–57
capitals, 104
casement windows, 91
cat's paw nail puller, 20, 40
cedar, 32
ceramic/porcelain
    about, 21, 81, 82
    basics of, 9
    see also stone
chandeliers, 105
chemical stripping, 44, 66, 108
cherry, 30
chestnut, American, 29
chicken wire glass, 96
circle sawn wood, 36
circuit tester, 20

circular saw, 20
claw hammer, 20, 41
clawfoot tubs, 103
clinkers, 83–84
closet doors, 53
coat rack, doorknob, 155–156
coffee table, pallet, 133–134
columns, 104
common bricks, 82–83
contractors, 23
copper, 63, 67, 68, 69, 109
corbels, 105
cordless drill bit assortments, 20
cornices, 66, 105
corrugated glass, 96
countertops, 72, 142–145
cypress, 32

debarked (DB), 58
deconstruction
    definition of, 8
    tools for, 20
demolition companies, 23
denailing reclaimed lumber, 40–41
denatured alcohol, 43
dents, 46
desoldering copper pipe, 69
divided lites, 90
doorknobs
    about, 102, 104
    cleaning reclaimed, 107
    coatrack with, 155–156
    gallery photos of, 16
doors
    about, 35
    "handedness" of, 49
    hanging reclaimed in existing jamb,
        122–125
    hardware for, 104
    island backed with, 126–132
    paneled, 49–51
    reclaiming wood, 48–49
    shoe for, 46
    types and styles, 49–53
double-hung windows, 90–91
Douglas fir, 32

eight-panel Colonial doors, 50
elm, antique, 30
enameled cast-iron tubs, refinishing,
    107
end grain flooring, 120–121
end-cut pliers, 20
extractor pliers, 20, 40
eye protection, 65
eyebrow windows, 92

face brick, 83
faceplates, 14
faucets, 103
fieldstone, 75, 80
finishes, wood, 42–44
firebricks, 84
fireplace surrounds, 72–73
fixed windows, 90
fixtures
    about, 101–102
    reclaiming plumbing, 106–107
    types of, 102–105
flagstone, 75, 80
flea markets, 23
floor tiles, 72
flooring
    about, 34
    end grain, 120–121
    laying with reclaimed tongue-and-
        groove planks, 115–118
    milling tongues and grooves for,
        112–113
    reclaiming vintage wood, 47–48
    routing tongues and grooves for, 114
    salvaging and laying pegged plank,
        119
four-panel traditional doors, 50
French doors, 52
frosted glass, 94
furniture, from pallets, 59, 133–134

gallery, 11–17
galvanized steel, 68
garage sales, 23
gargoyles, 105
glass
    about, 21, 87–88
    basics of, 9
    cutting reclaimed, 99
    gallery photos of, 13
    types of, 93–94
    vintage specialty, 95–96
    see also windows
gouges, 46
granite, 74, 77
granite-slab kitchen island, 142–145
greenhouse, old-window, 146–151

hand hewn wood, 37
hand maul, 20
handles, 14
hardware
    about, 104
    cleaning reclaimed, 107
    polishing and refinishing, 108–109
    removing paint from, 108

hardware, plumbing, 103
heartwood, 28
heat stripping, 43–44
heat treatment (HT), 58
hemlock, 33
hickory, 30
hinges, 104, 107
hooks, pipe, 139
hopper windows, 91
horizontal raised panel doors, 50
horizontal surfaces, stone, 73

industrial glass, 96
IPPC (International Plant Protection
    Convention), 58
iron, 64, 68, 109
island
    door-backed, 126–132
    granite-slab kitchen, 142–145

jalousie windows, 91

kiln-fired materials, 80–82. see also
    bricks; ceramic/porcelain
kitchen sinks, 102–103

lacquer thinner, 43
laminated lumber, separating, 41
landfills, 8
laser level, 20
lead paint, 24, 43, 44, 121
lead pipes, 68
leaded windows, 93
lighting fixtures, 13, 105, 109
lime, 107
lime-based mortar, 84
limestone, 74
lintels, 105
locking-jaw pliers, 20
locksets, 104
lumber
    denailing, 40–41
    salvaging, 34–35
    separating laminated, 41
    see also wood

machined glass, 95
mail slots, 104
mantels, 10, 11, 54, 72
maple, 30
marble, 75, 77
masonry, about, 21. <I>see also<m>
    stone
metal
    about, 21, 61–62
    basics of, 9

gallery photos of, 15
    reclaimed tin ceilings, 64–67
    types of, 63–64
metal detector, 20, 40
methyl bromide (MB), 58
methylene chloride, 66
mineral oil, 43
mirrors, 88, 89
mortar, removing, 84–85
muriatic acid, 85

nail holes, 46
nippers, 40

oak
    antique, 31
    white, 31
online resources, 23

paint
    about, 42
    removing from hardware, 108
    stripping, 43–44
    painted windows, 93–94
    pallets
    projects for, 133–135
    reclaiming and using, 57–59
panel lites, 52
paneled doors, 49–51
pathways, 73
pavers, 21, 73, 80, 83
paving brick, 83
pedestal tubs, 103
pegged plank flooring, 119
penetrating oil, 42
pick axe, 20
pine, 33
pipe
    coat hooks from, 139
    salvaged plumbing, 67–69
planing, 55
planks, 34
planter, pallet, 135
plumbing fixtures, 102–103, 106–107
plumbing hardware, 103
plumbing pipe
    coat hooks from, 139
    salvaged, 67–69
    towel rack from, 140–141
pocket doors, 53
polyurethane, 42, 43
portal windows, 92
posts, 104
pry bars, 20
pumice stones, 106

quarter-sawn wood, 37

raised-panel doors, 50, 51
recessed-panel doors, 51
reciprocating saw, 20
reclaimed materials
    assessing, 24–25
    benefits of, 8–10
    choosing among, 20–21
    gallery of, 11–17
    sourcing, 22–23
    see also individual materials
redwood, 33
resilvering mirrors, 89
resources, 157
respirator, 24, 65
resurfacing salvaged timbers, 55
retail salvage companies, 22
rot, wood, 45
rough sawn wood, 36
routing tongue-and-groove
    flooring, 114
rust stains, 106
safety, 22, 24, 58, 63, 65, 67, 69, 78, 121
safety lights, 105
salvage companies, 22
sandblasted glass, 94
sanding, 44, 55
sandstone, 74
scale, 107
scraping paint, 44
screw holes, 46
screws, replacing nails with, 57
sealants, stone, 76
shellac, 42, 43
shoe, for doors, 46
showerheads, 103
siding, 39, 47
single-hung windows, 90–91
sinks, 102–103
six-panel Victorian doors, 50
skip sawn wood, 36
slate, 75, 79
sledge hammer, 20
smooth planed wood, 37
specialty glasses, 96
spigots, 103
sprayed-on coatings, 24
square blade shovel, 20
stained glass, 93–94
stainless steel, 64
statuary, 105
steel, 109
stone
    about, 21, 71–72
    basics of, 9

brick, 82–85
  cleaning, 76
  cutting, 77–79
  gallery photos of, 12, 13
  kiln-fired materials, 80–82
  loose, 80
  sealing, 76
  tiles and slabs, 76
  types of, 74–75
  uses for, 72–73
  working with, 76
stripping paint, 43–44
structural timbers, reclaiming vs.
  buying, 125
stud sensor, 20
swap meets, 23

terra-cotta tiles, 80, 81–82
textured glass, 95
timbers, 34, 53–55, 125
tin
  about, 63
  backsplash, installing, 136–138
  ceilings, 64–67
  tobacco barn wood, 35

toilets, 103
tongue-and-groove flooring
  laying, 115–118
  milling, 112–113
  routing, 114
towel rack, plumbing pipe, 140–141
trisodium phosphate, 106
tubs, 103, 107

urea formaldehyde, 24
utility knife, 20
utility sinks, 102–103

varnish, 42, 43
voltmeter, 20
wall sconces, 105
walnut, 31
wax, 42, 43
wet saws, 77–78
white oak, 31
wholesale salvage companies, 22
windows
  about, 88
  building greenhouse with, 146–151
  cabinet from, 152–154

reviving reclaimed, 97–98
  types of, 90–94
  see also glass
wood
  about, 21, 27–28
  basics of, 9
  cabinetry, 56–57
  denailing reclaimed lumber, 40–41
  doors, 48–53
  finishes on, 42–44
  flooring, 47–48
  gallery photos of, 11, 12, 15, 16, 17
  grades for, 38
  moisture content and, 41, 48
  pallets, 57–59
  reclaiming, 38–41
  repairs for, 45–46
  salvaging lumber, 34–35
  species of, 29–33
  textures of, 36–37
  tobacco barn, 35
  trim and siding, 39
wood consolidation products, 45
wood filler, 46
wood trim, 39

PHOTOGRAPHY CREDITS